THE BIG QUESTIONS
Mathematics

Tony Crilly is Emeritus Reader in Mathematical Sciences at Middlesex University, having previously taught at the University of Michigan, the City University in Hong Kong and the Open University. His principal research interest is the history of mathematics, and he has written and edited many works on fractals, chaos and computing. He is the author of the acclaimed biography of the English mathematician Arthur Cayley and the internationally bestselling *50 Mathematical Ideas You Really Need to Know*.

The Big Questions confronts the fundamental problems of science and philosophy that have perplexed enquiring minds throughout history, and provides and explains the answers of our greatest thinkers. This ambitious series is a unique, accessible and concise distillation of humanity's best ideas.

Series editor **Simon Blackburn** is Professor of Philosophy at the University of Cambridge, Research Professor of Philosophy at the University of North Carolina and one of the most distinguished philosophers of our day.

Titles in *The Big Questions* series include:

PHILOSOPHY
PHYSICS
THE UNIVERSE
MATHEMATICS

THE BIG QUESTIONS

Mathematics
Tony Crilly

SERIES EDITOR
Simon Blackburn

Quercus

Contents

Introduction

Mathematics is something we should all know about. The school syllabus is one thing – and does not excite all – but the subject offers much more. A silent partner in scientific applications, it also has fundamental connections with the arts. As part of the human heritage, mathematics is alive and constantly expanding its boundaries; its life is sustained by the 'Big Questions'.

The big questions in mathematics are very varied. Some are initiated by the seismic changes in modern technology, while others originated in Ancient times and reverberate to this day. Some have received a definite answer only to be replaced by new batches of questions, but others have persisted, never retiring, even after centuries on the frontline. Those verging on the philosophical might never be resolved one way or another, but the questions remain fascinating nonetheless.

Such is the way of mathematics. A curious fact is that mathematics advances slowly. While there is a premium on speed in mental arithmetic and tricky little problems at school there is absolutely no advantage gained by being fast in the real business of mathematics. That mathematics does advance is undeniable, but its progress resembles the gradual inevitability of a lava flow more than the 'Eureka!' moment of a great genius.

Mathematics has a distinctive nature that separates it from science. When a scientific theory loses credibility – like the once-popular 'phlogiston' to explain why objects burn, or the 'luminiferous aether' to explain the transmission of light – it is abandoned. Such theories are past their sell-by dates and placed in science's history book for antiquarian interest only. In mathematics, things are different. A proven result cannot subsequently be proved false, and so a theorem – a proven mathematical fact – has an infinite life. Pythagoras's theorem about right-angled triangles is true for all time.

Mathematicians today may not write research papers to contribute theorems of the type that Euclid was writing in 300 BC. Yet these works can inspire, and new ways of thinking can be discovered from foundational texts. We may read the Greek mathematician Diophantus on the theory of equations and still learn from it, for some types of equations in Ancient Greece remain unsolved to this day.

This is not to say that time has no effect on mathematical theory and theorems. They are often modified, refined and tailored for a modern context. The tendency in mathematics is for results to be swallowed up by generalization, and their eventual fate is not the dustbin but a footnote to a more general theory.

We live today in exciting times for mathematics. New questions have to take account of the computer age. This is not merely because computers are efficient at adding up columns of figures, but because they challenge our notions of mathematical proof and raise questions on the nature of mathematics. They can deal with algebra and show geometrical shapes and surfaces to advantage.

The Big Questions considered here focus on big issues and tackle the basic questions that need to be addressed. They will show where mathematics has come from, where it has travelled and where it might be going. They all yield answers, though by no means are those answers all cut and dried. They raise the problems that excite mathematicians and tell us how mathematics informs us about the real physical world we live in. And most of all, they demonstrate that mathematics is a living and breathing subject.

WHAT IS MATHEMATICS FOR?
An introduction to purposes and prospects

*I*n the 21st century, mathematics is a vast and multifaceted *subject. It covers such a broad spectrum of activity that it appears scarcely possible to classify all its manifestations within a single subject. At one end of the spectrum, it defines the nuts and bolts of counting, time and money that enable daily life to chug away. At the other end, it can seem a sealed world in which great ivory-towered minds manufacture puzzles of mammoth complexity – which they then devote years to trying to solve. At the same time, our politicians consistently tell us we need* more *mathematicians. What, then, is all this mathematics for, and how does it fit into our world?*

The mathematics we live with today had its seeds in early numerical culture traceable to around 3000 BC. Unsurprisingly, the beginnings were geared to dealing with practical matters. Problems of the market place, the payment of taxes, measuring one's land, comprehending the stars and the planets, devising a calendar – all were applications requiring numbers, calculation and some rudimentary geometry. But with the Egyptians, a thousand years later, societies began to investigate the properties of their number systems irrespective of obvious applications. They also began to create, out of curiosity and intellectual pleasure, mathematical puzzles, just as we might enjoy the Sudoku page in the newspaper. Mathematics had begun to look to itself. The mathematician was born.

Colossal strides were taken by the Ancient Greeks around 500 BC, when a true culture of mathematical thought flourished. The works they produced have been influential down the ages and are still studied today. Mathematics was regarded as being of the highest good and formed an intrinsic part of the classical education. Pythagoras, Plato, Archimedes, Euclid are just some of the Greek philosophers who championed mathematics and who exerted an influence for hundreds, even thousands, of years afterwards.

During the first centuries of Christianity, the pendulum swung back, and those who were mathematically inclined could find themselves cast out to the fringes of the cultural world. Around AD 400 St Augustine of Hippo suggested that 'the good Christian should beware of mathematicians and all those who make empty prophecies', condemning them for making 'a covenant with the devil to darken the spirit and to confine man in the bonds of Hell'. In those days, mathematicians were closely connected with the murky practices of astrologers, and suspicion about potentially nefarious or heretical purposes tended to hang over mathematics for a long period.

In the 16th century, the philosopher Francis Bacon lamented the fact that 'the excellent use of the pure Mathematics' was not well understood, but a sign of better things was that Galileo took up his position as professor of mathematics at the University of Padua. Galileo's encounter with the Catholic Church, which rejected some of his findings, showed that tolerance of mathematics, and its implications for physics and astronomy, had limitations. But the later 17th century unleashed a mathematical and scientific revolution, in the shape of Isaac Newton and his contemporaries, which would forever change the cultural balance of power. The Romantics of the late 18th and early 19th centuries might decry these new worldviews, and William Blake might satirize Newton, but with mathematics as the language of science its future was secure. The 19th century saw the establishment of mathematics in universities everywhere and witnessed a flood of new and challenging work. Mathematics was here to stay.

Practicality and purity

There is a popular debate about mathematics, about whether necessity is the mother of mathematical invention or whether innovative mathematics creates opportunities for application. Historically, practical considerations were the drivers of mathematics, but once the internal life of the subject opened up there was the possibility that 'pure' mathematical thinking could itself create the space for new applications. Good mathematics is rarely removed from potential application, but one never knows when that moment of application might come. A sharp insight might be taken up next week or it might lie dormant for 50 or 500 years.

History is strewn with examples of the purely mathematical theory finding its practical partner. The Ancient Greeks elaborated a theory of conic sections, and this proved to be just what was needed in the 17th century when Johannes Kepler and Isaac Newton asserted that the planets moved in ellipses. 'Matrix algebra', the theory of multi–dimensional numbers, was elaborated in the 1850s to deal with internal mathematical problems; it was precisely what was needed for 'matrix mechanics' in the fast-moving quantum theory of 70 years later. And when George Boole set up a system for turning logic into algebra, giving us 'Boolean algebra', he was not to know that he was furnishing the language for computer programming a century later.

Only 50 years ago, the influential English mathematician G.H. Hardy wrote that he pursued mathematics unconstrained by any thought of having to make his ideas of 'practical relevance'. Indeed, he took comfort from the theory of numbers being remote from practical applications. He would not be able to celebrate its insularity today, not in a world where his kind of pure mathematics is of the utmost importance in terms of computer security (see *Can We Create an Unbreakable Code?* and *Is There Anything Left to Solve?*). Today we have many theories of dimension, but when Benoît Mandelbrot drew attention to 'fractals' in the 1970s, few would have guessed at their potential applicability (see *Why Are Three Dimensions Not Enough?*).

But mathematicians do respond to needs as well. In the 18th century, James Watt had a problem turning the linear motion of a piston in his steam engine into rotary motion, with the result that the theory of geometrical linkages had its birth during the Industrial Revolution. When codebreakers were needed during the Second World War (see *Can We Create an Unbreakable Code?*), mathematicians were recruited from universities for their special skills, and the result was the construction of the world's first electronic computer.

Thus, pure mathematics and applied mathematics continue a symbiotic relationship, and never was this more true than in the electronic age. Without mathematics, computers would be useless, digital photography would be impossible and mobile phones would fall silent. But the professional mathematician's 'pure' research is also now significantly powered by the computational ability of computers: the 'applied' feeds the 'pure' in turn.

Mathematics has its self-conscious side too, its philosophically reflective side. The history of this shows a movement away from the Ancient Greek assumption that mathematicians unearthed pre-existing truths to a much more finely nuanced conception of the mathematician's role, in which creativity and imagination are involved (see *Is Mathematics True?*).

In modern mathematics, the way of proceeding is based on axioms and logical deduction. The Greeks assumed the truth of their axioms, but today's mathematicians expect only that axioms be consistent. In the 1930s Kurt Gödel rocked mathematics when he proved his 'incompleteness theorems', which held that there were some mathematical statements in a formal axiomatic system that could neither be proved nor disproved using only the axioms of the system. In other words, mathematics could now contain unprovable truths that might just have to stay that way.

Varied and vast though the modern mathematics may be, at its root lies the school-curriculum division into arithmetic, algebra and geometry. What lies at their core, and where are they going?

Numbers and their properties

The numbers used for counting remain the most important in the mathematical repertoire; they are where a mathematician starts. The history of their evolution (see *Where Do Numbers Come From?*) is a rich one, and it is certainly not inevitable that we ended up with a 'base ten' system using the symbols 0–9. For a start, at first there was no zero.

The properties of prime numbers – numbers which can only be divided by themselves and the number 1 to produce another whole number – are particular objects of fascination. Surprisingly, there are many things unknown about them. We still do not know how they are distributed among the counting numbers, which may be difficult to believe since we have known about prime numbers for more than 2000 years (See *Why Are Primes the Atoms of Mathematics?* and *Is There Anything Left to Solve?*). Beyond the counting numbers and those of them that are primes, the repertoire has expanded over the centuries to embrace negative numbers, fractions and then the so-called 'irrational numbers' of infinitely receding decimal places without pattern. All of these, together, mathematicians call the 'real' numbers (see *Which Are the Strangest Numbers?*).

That was not all. The 'real' numbers are all one-dimensional. They can be conceived of as spreading left (negative numbers) and right (positive numbers) on a number line. A great leap forward came when mathematicians ventured into two dimensions with what they called 'complex numbers' (see *Are Imaginary Numbers Truly Imaginary?*). These delivered mathematicians greater power to solve equations and offered new theories of analysis. Today, 'complex' numbers are indispensable in the study of phenomena such as electricity and magnetism.

There are, then, many types of numbers, but where do they end? Mathematicians from the earliest times had to grapple with the issue of infinity. It was assumed, from Aristotle onwards, that there was 'potential infinity' – one infinity, which could never be reached. But in the 19th century Georg Cantor introduced another notion of infinity, and it became possible to talk of *many* infinities (see *How Big is Infinity?*).

Geometry, algebra and mathematical revolutions

For millennia, geometry was in thrall to the Ancient Greeks and what appeared to be their undeniable authority, which laid down many of the rules that schoolchildren absorb to this day. In particular, Euclid built up a body of geometrical knowledge built on his cast-iron logic and presented as the canonical truth. But, over time, cracks began to appear in Euclid's geometry, and eventually it became clear that there were *other* valid geometries that dealt with phenomena in two, three – and more – dimensions (see *Where Do Parallel Lines Meet?*) and which have resulted in the concept of the 'manifold' – a shape that has different local and global geometry (see *What Shape Is the Universe?*). These geometries may even have a greater claim than Euclid's to be the 'geometry of the universe', the subject that is so compelling for physicists.

While physicists appropriate geometry to hunt down the secrets of matter and the universe, biologists and medical researchers take a different type of geometry, 'knot theory', to attempt to untangle and analyse DNA – a practice that has yielded the forensics of DNA profiling, and which has had significant ramifications for issues of human identity and the solving of crimes. All in all, mathematicians have provided scientists with different geometries as a kind of toolkit from which they can select what seems right for the particular job at hand.

There comes a point when geometry translates into the language of algebra, a development credited to Descartes in the 17th century. The 20th century, then, saw the geometry of symmetry metamorphose into algebra too. Symmetry, the elusive property that has often been taken in mathematics – as in much else – to define beauty (see *Is Mathematics Beautiful?*), can now be captured mathematically by 'group theory'. Groups lie at the centre of modern algebra and give a means whereby symmetry can be examined on a microscopic scale (see *What Is Symmetry?*). In a huge research project, whose beginnings stretch back to the 19th century, mathematicians eventually completed the classification of finite groups in 1981. In what became the 'enormous theorem', a map of the groups was created in which groups fell into known families plus 26 sporadic groups, the largest of the latter containing approximately 8×10^{53} members – that's 8 followed by 53 zeros. Group theory occupies an important place now in theoretical physics, where transformations of space form groups, and in chemistry and crystallography too, where symmetries come into play.

'Finding the value of x' in an algebraic problem is something every school-level mathematician becomes familiar with. These types of 'inverse' problems are an area where mathematics excels, with applications far and wide. In these, we often need to find an 'unknown' but at first we can only find a relationship or an equation that involves the unknown. If we are told, for example, that increasing the sides of a square field by 3 metres will result in a field of 400 square metres, we can work out the unknown length x of the original field as an inverse problem. Using algebra and 'unwrapping' the equation $(x + 3)^2 = 400$ gives us $x = 17$. When the work of previous generations of mathematicians has yielded us an array of formulae to do these tasks, we can take welcome shortcuts (see *Is There a Formula for Everything?*).

Launching a rocket into space involves 'differential' equations, and this means the apparatus of 'the Calculus' (see *What Is the Mathematics of the Universe?*), a method typically used to measure rates of speed and acceleration. There are specific types of differential equations, supported by a well backed-up theory, but there are also many 'one-off' equations that defy exact solutions. Henri Poincaré established a new branch of the theory of differential equations as a 'qualitative theory', which focused on the *properties* of solutions rather than finding the solutions explicitly. This study gave rise to the theory of 'chaos' (see *Can a Butterfly's Wings Really Cause a Hurricane?*) and gave a distinctive orientation to the new theory of topology, a radical departure in the way we looked at shapes (see *What Shape Is the Universe?*).

The new and unknown mathematics

'Topology' might not trip off the tongue of the average non-mathematician, but two other relatively late developers are far more familiar terms: probability and statistics.

One of the outstanding modern creations of mathematics, probability theory (see *Can Mathematics Guarantee Riches?*), enables us to handle uncertainty in a quantitative way. The recreational mathematics of the 17th century gave us the beginnings of this theory, in the analysis of gambling problems, and now, smoothed out and developed into a rigorous calculus, it is the backbone of the analysis of risk. Statistics, a related field (see *Are Statistics Lies?*), provides the theory for handling data properly and the context for carrying out experiments. Statistics had some beginnings in agricultural experiments, but its methods are now used so widely that there is scarcely a part of human activity, from politics to medicine, that is statistic-free.

Using the results of statistics and other mathematics leads naturally to the desire to make predictions, to know the future (see *Can Mathematics Predict the Future?*). The demographer wants to make a reasonable prediction of the population in five years' time. The trader will try to second-guess the stock market on the basis of statistical evidence and hunches. How is this to be done? These are difficult questions, as is the business of weather forecasting, which depends on mathematical equations that, as yet, cannot be solved (see *Is There Anything Left to Solve?*), and whose difficulty is compounded by the 'butterfly effect' (see *Can a Butterfly's Wings Really Cause a Hurricane?*).

So, there is old mathematics and there is new mathematics. Lest we sit back and think the job is nearly done, we should remind ourselves that there is also unsolved mathematics, and lots of it (see *Is There Anything Left to Solve?*). And just as well, for if that were not the case, mathematics would wither on the vine. There are some great unsolved questions that have baffled thinkers for years, such as the Goldbach conjecture and the Riemann hypothesis, both of which are related to prime numbers; there are also some noisy new problems. There has been progress, of course, and some of it headline-grabbing. Mathematics leaped into the public gaze with the solving of Fermat's last theorem in 1994 (see *Is Mathematics Beautiful?*). Before that, mathematics and computing joined forces to solve the 'four colour' theorem (see *Is There a Formula for Everything?*), and recently a reclusive Russian mathematician stunned the world by proving the hundred-year-old Poincaré conjecture – and not even claiming his £1 million reward.

What, then, is mathematics for? In some ways this is an odd question. We tend not to ask 'What is music *for*?' or 'What is literature *for*?' We accept that they simply are activities and thought processes and exercises of the imagination in which human beings indulge – they always have, and they always will, because they must. If one wants to look for applications, then they are all around us and proliferating. If one wants to advance all the ways in which mathematics imparts knowledge of the world, of the universe, of nature and of human interactions, one can do that too. There is an inestimable amount that mathematicians can *do*, and have *done*, in ways that change lives. But, at its root, mathematics is motivated by that basic, defining feature of humankind – insatiable curiosity.

WHERE DO NUMBERS COME FROM?

From notches on bones to hexadecimals

*I*n *our daily lives we are immersed in numbers. We wake and blearily absorb the circle of numbers on the clock face, or more likely the luminous glow of a digital alarm; we may rush to college on the number 134 bus, or race to work on the 08.32 train to Paddington; we count out change to buy our lunch, we check our diary dates, we punch the buttons on our mobile phones; at the end of the day, we might idly scroll through the dizzying array of numbered TV channels, until finally we get to bed, with a last glance to check the time. So deeply embedded are numbers in our lives, and we in the world of numbers, that we do not stand back to see them for the astonishingly versatile things they really are. So where did they come from?*

Of course, many of the numbers we encounter are simply designations, labels. Bus routes could, in theory, be evoked by means other than by numbers. The famous Heinz '57 Varieties' or Jack Daniels 'Old No. 7' bourbon slyly suggest a series of subtly differentiated canned foods and whiskies, but they too are simply invented labels – brands. Even these brands depend for their effectiveness on the ways in which human societies have elaborated systems of numbers for the purposes of ordering objects – first, second, etc. – and counting them.

Today, almost universally, the human race adopts the same system of just ten symbols, from 0 to 9, to do its counting and ordering. The combinations of those symbols are versatile enough to express the vast distances of galaxies down to the diameter of an atomic nucleus, and to express them in different ways. The layman might write that the Earth is 93,000,000, or 93 million, miles from the Sun, whereas a mathematician or scientist would likely prefer the concise elegance of 9.3×10^7 (that is, 9.3 multiplied by 10 to the power of 7). For describing a thousandth part of a metre, we have three options: 0.001 metres, 1 millimetre and 10^{-3} metres (that is, 10 to the power of -3), while the tiny diameter of a nucleus can be neatly expressed as 10^{-15} metres.

Elegant and versatile though those ten symbols are, they were never inevitable, and they were neither the beginning – nor the end – of mankind's invention when it comes to number systems.

The earliest counting methods

Researchers have found evidence as far back as 30,000 years ago of the early recording of numbers on 'tally sticks'– sticks of wood carrying marks signifying quantities. In Africa and Eastern Europe, notched bones used for numerical records have provided further evidence of practical engagement with the act of counting. Tally sticks were still being used in England for the purposes of tax collection in the 13th century, and surprisingly this traditional system lasted until the 1820s, when paper records took over. Today, tally marks still serve a purpose when the aim is to record a continually rising quantity, such as the number of points scored in a game or data for statistical investigations. The familiar method involves counting to five to form an inscription resembling a barred gate, and it seems to have very early origins.

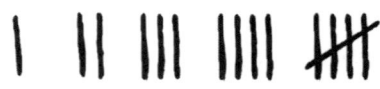

THE BARRED–GATE TALLY SYSTEM

A SOUTH AMERICAN TALLY SCHEME

Tally methods are found all over the world. One scheme, from South America, also adopts a system of five lines, albeit producing a different shape.

Among the aboriginal peoples of Australia, primitive counting systems generalized beyond small numbers. A counting system translating as 'one', 'two' and 'many' was used in Tasmania, while on the mainland, in Queensland, an instance of 'one', 'two', 'one and two', 'two twos' and 'much' has been recorded.

Babylonians and Egyptians

The emergence of a proper number system occurred in the 'cradle of civilization', the Middle East. The Babylonian civilization flourished in Mesopotamia, that part of modern Iraq that lies between the Tigris and Euphrates rivers; its capital, Babylon, sat 50 miles to the south of modern Baghdad. In the 3rd millennium BC, the Babylonians were using a number system that revolved around the number 60. Vestiges remain in our culture today – in our measurement of time (60 seconds in a minute, 60 minutes in an hour) and, because the Babylonians used mathematics in their astronomy, in the number of degrees in a circle or full rotation: 360. A system based on the number 60 has mathematical merits, one of which is that 60 can be divided by no less than eleven smaller numbers, 1, 2, 3, 4, 5, 6, 10, 12, 15, 20 and 30, which yields obvious benefits for the apportioning of quantities.

The numbers 1–60 were denoted by just two symbols that were easy to inscribe on the clay tablets used for recording them: a vertical single line and a wedge. The number 1 was the line symbol and the numbers 2–59 were denoted by different combinations of the two symbols. The Babylonians read their notation from left to right, and the number was defined by the relative position of symbols in the row: in other words, the same as our system today. When they got to 60 they started again and used the same vertical single-line symbol for number 60 as they did for number 1: the only way to interpret it would be by context, so that if angles were being measured it would more likely be 60 degrees than 1 degree. They did not have an equivalent for zero.

THE TWO BABYLONIAN NUMBER SYMBOLS

THE NUMBER 23 IN BABYLONIAN SYMBOLS

The Ancient Egyptians developed a different and sophisticated system, expressed amply in their extraordinary

9, ∩

construction of the pyramids, a feat that required a knowledge of three-dimensional geometry and fairly accurate measurement. From around 2700 BC the system emerged, based on the number 10. For the numbers up to 9 they used vertical-line tally marks, then two different symbols for numbers 10 and 100. Unlike the Babylonians, the Egyptians wrote from right to left. The symbols for larger numbers were ornate: for example, a bird represented the number 100,000.

THE NUMBER 234 IN ANCIENT EGYPTIAN SYMBOLS

The Egyptians' mathematics was tied principally to practical matters, but they possessed some elaborate arithmetical tricks. Their scheme for multiplication was ingenious. In our modern system, mental multiplication requires our knowing our multiplication tables, often taught by rote in school. But children living in the early Egyptian times would effectively need just the two-times table, for they approached multiplication by a method geared to the use of the abacus.

The Vedic Aryans

Further east, during the second millennium BC the Vedic Aryan civilization spread from Central Asia into India's Indus Valley, and records of its method of arithmetic are dated to around 1000 BC. In this culture 19 mathematical 'sutras', or word formulae, were found among the poetry, literature and wisdom contained in the ancient Hindu texts known as the *Vedas*. The sutras provided shortcuts for, or alternative ways of, addressing a range of arithmetical problems. One, the 'vertical and cross-wise' sutra, for example, helps with the multiplication of whole numbers – a lengthy mental calculation in our modern number system. It works by a combination of multiplication and addition. If we wanted to multiply 13 by 24 – let's assume we're using our familiar symbols – we first write 13 over 24 in a grid.

1	3
2	4

A VEDIC ARYAN NUMBER GRID

The 'vertical' numbers are multiplied and then aligned to produce the number 212: that is $(1 \times 2) + (3 \times 4) \rightarrow 2 + 12$ makes 212. Then the diagonal, 'cross-wise' terms, added together, give the number of tens: that is $(1 \times 4) + (3 \times 2) \rightarrow 4 + 6 = 10$ which makes 100.

Now, adding 212 and 100 produces the correct result of 312. The method may appear to be a sleight of hand, but its validity lies in the principle that to multiply two numbers, say $ab \times cd$, actually requires the multiplication of $(10a + b) \times (10c + d)$.

The advent of zero

Zero was a latecomer as a mathematical symbol. It has two main functions, and the first to manifest itself was as a placeholder, a way to distinguish (for example) the number 27 from the number 207. Neither the Babylonians, the Egyptians, the Greeks, nor the Romans had a symbol for it – they had nothing for expressing 'nothing'. The second function of zero is as a genuine number, and here the origins go back to the Indian mathematician Brahmagupta, who attempted to integrate it into the number system around AD 600.

The Latin name for zero (*cifra*) evolved into *zefro* and in Italian became *zero*. In French, *chiffre* ('zero') was translated as 'cipher' in English, a usage now obsolete. In English we use 'nought' to mean zero, a derivation from the word 'nothing'. Nought is really a mathematical misnomer, for it would be wrong to think of the number 0 as nothing. It is certainly something, a symbol that gives our modern notation its ultimate power. As Brahmagupta realized, the mathematical challenge was to harmonize the interloper number 0 with the rest of the number system.

Number bases and the vigesimal legacy

Mathematicians speak of underlying number 'bases' for the counting systems of different cultures. They may be understood as the core units, or building blocks, of a counting system. Our modern base is ten – the decimal system, reflected in its ten symbols 0–9. But history has witnessed cultures that, adapting numbers to their needs, have adopted bases of 2, 3, 4, 5, 12, 20, and of course the Babylonians' 60. There are traces of that mixed heritage in our system even now, as evidenced in the language of numbers. In German and English, for instance, the words 'eleven' (German *elf*) and 'twelve' (German *zwölf*) appear linguistic oddities, forming a break between the numbers up to ten and the subsequent 'teen' numbers that derive from the word 'ten'. They are a legacy of a base 12 system, as indeed is the pre-decimal quantity of 12 pence in a shilling.

In what is today Guatemala and south-eastern Mexico, the pre-Conquest Mayan civilization adopted a base 20 number system, which we refer to as a 'vigesimal' system. (Interestingly, the Mayans were also one of the first peoples to formulate a notation for numbers based on a positional system that embraced a concept of zero.) The roots of such systems are often traced to our ten fingers and ten toes, and many cultures contain legacies of base 20 systems. In English (and its French and German counterparts), the 'teen' sequence of numbers breaks off at nineteen to make way for the special word 'twenty'.

> '*The creator of the universe works in mysterious ways. But he uses a base ten counting system and likes round numbers.*'
>
> SCOTT ADAMS,
> AMERICAN CARTOONIST

The quantity 20 imbues our language and culture in further ways. It has a synonym in the word 'score', a word most memorably invoked in the Book of Common Prayer's description of a 70-year lifespan as 'threescore years and ten'. English imperial measures include the hundredweight (cwt), and it takes 20 cwt to make a ton. And, of course, before decimalization of pounds sterling (in 1971), there were 20 shillings to every £1. The French language contains a remnant of the base 20 system in its unusual word for eighty – *quatre-vingt* ('four-twenty').

The decimal system
Despite this multifarious legacy of number systems, the world today has coalesced around the base ten 'decimal' system.

In many ways, it is a natural human choice: we have ten fingers on which to count. The Ancient Romans operated a base ten system for counting with whole numbers but their system of fractions was based on 12 for easier handling, since 12 is divisible by 2, 3, 4 and 6. (Some historians have accounted for 12 as a base because we have three joints in each finger, making 12 on each hand, here excluding thumbs.) In truth, the Roman contribution to the development of mathematics was minimal. But the language of Latin evolved into many of the modern European languages, so Latin has largely given us our *words* for numbers; and the Roman system for writing numbers, I, II, III, IV, V, …, X, etc., endures in parallel with the 'ordinary' digits we use, particularly in the recording of dates.

The number symbols we use today are those used first by Indian mathematicians of Brahmagupta's era and inherited by scholars in Arabia. The symbols were then transmitted by Arab travellers, traders and conquerors as they spread out along North Africa and into the Iberian Peninsula. Arab mathematical scholarship began to spread in the West in the 12th century too: the ninth-century mathematician al-Khwarizmi's *On the Calculation with Hindi Numerals* appeared in a Latin translation, and one Leonardo of Pisa (also known as Leonardo Fibonacci) promulgated this Hindu–Arabic numbering system in his *Liber Abaci* ('The Book of Calculations'), published in 1202. In the 13th century, the English philosopher, mathematician and heretical monk Roger Bacon recorded these symbols.

$$1\text{7 3}\text{8 4 6}\text{A 8 9 }10$$

ROGER BACON'S NUMBER SYMBOLS (13TH CENTURY)

With just a few changes, by the 16th century the ten symbols in widespread use resembled the ones we use today, having achieved greater standardization following the advent of mass printing.

Binary numbers

While the decimal system has proved powerful and resilient, modern computer technology has generated a need for a different kind of system. Every 'switch' in a computer is either 'on' or 'off', so computer technology is based on a recognition of just these two modes. This has produced the binary system, with the two symbols 0 and 1 as its alphabet. Since we do not naturally think in terms of binary numbers, we need a technique for converting decimal numbers into binary numbers, and vice versa.

Decimal number	Binary number
0	0
1	1
2	10
3	11
4	100
5	101
6	110
7	111
8	1000
9	1001

DECIMAL–BINARY CONVERSION CHART

In the decimal system we think in terms of powers of 10 – thus the number 312 is made up of 3 hundreds, 1 ten and 2 ones. For the binary system we must think rather in terms of powers of 2 – in terms of doubling: 1, 2, 4, 8, 16, 32, 64, 128, 256, 512, etc.

Therefore, the conversion of the number 312 into a binary number involves dividing, beginning with the highest power of 2 that will fit into 312, which is 256, and then moving down the scale. Thus we find that 312 = 256 + 32 + 16 + 8. But then, to obtain our binary number, we must also take account of those powers of 2 that are *not* represented in this expression, allocating a '1' to the powers that are present and a '0' to those that are absent. In full, that produces:

$$312 = 1 \times 256 + 0 \times 128 + 0 \times 64 + 1 \times 32 + 1 \times 16 + 1 \times 8 + 0 \times 4 + 0 \times 2 + 0 \times 1$$

In other words, the decimal number 312 produces the binary number 100111000, or

$$312_{10} = 100111000_2$$

using subscript numbers to indicate the base of the number.

There is, however, a neat two-stage trick for converting a decimal number into a binary one, similar to the doubling method the Ancient Egyptians used for multiplying. First, we place the number to be converted on the far right-hand side of a grid and then divide the number by 2 until we arrive at the number 1, all the time *not bothering about any remainder*. So, taking 312 again as our example, at one point we will need to divide 39 by 2; we would record 19 as the answer, and ignore the remaining 1. This gives the top row of the grid. Second, in the bottom row of the grid, we record which numbers in the top row are odd and even, tagging every odd number as a '1' and every even number as a '0'. The sequence on the bottom row produces the binary number:

Decimal number	1	2	4	9	19	39	78	156	312
Binary number	1	0	0	1	1	1	0	0	0

DIVISION GRID FOR BINARY NUMBERS

Octal and higher systems

Unsurprisingly, one problem with binary numbers is that they can become very long sequences of ones and zeros, and thus use up valuable computer memory. We can, in theory, adopt any base we wish to, so a way to shorten binary numbers is to convert them using base eight arithmetic. In this 'octal' arithmetic we need eight symbols in our character set: 0, 1, 2, 3, 4, 5, 6 and 7.

The next base after eight capable of condensing even greater binary numbers into shorter expressions is the hexadecimal system, with base 16. A convenient character set for it consists of the 16 symbols 0, 1, 2, 3, 4, 5, 6, 7, 8, 9, A, B, C, D, E, F, where A corresponds to 10, B to 11, C to 12, and so on. To convert a binary number into hexadecimal we bracket it in fours, so that (starting from the right), for example:

$$312_{10} = (0001)(0011)(1000)_2$$

Again referring to the decimal–binary chart (and ignoring redundant zeros), we obtain the result:

$$312_{10} = 138_{16}$$

But it is possible to have a hexadecimal number composed entirely of letters, for example the decimal number 2748 would, in hexadecimals, become:

$$2748_{10} = 10 \times 16^2 + 11 \times 16 + 12 = ABC_{16}$$

The story does not end with hexadecimals. The power of computers marches on, and we are now familiar with base 32 and base 64 systems.

The long journey of numbers

The distance travelled by numbers, from primitive marks on bones to hexadecimals and beyond, has been immense. The journey has been spurred, in the first instance, by the practical needs of human societies. But it was not *just* the utility of numbers that concerned the ancients. Around 500 BC, while the citizens of Classical Greece had their number system for the market place, in what they termed *logistic*, thinkers such as Pythagoras and Plato were already developing their *arithmetic*, the theory of numbers. The independent fascination and beauty of mathematics asserted itself early on.

Two-and-a-half millennia later, the Hindu–Arabic decimal system has proved the most durable and widespread for the world's mathematical, scientific and daily needs. But the history of numbers also reveals how, at different times and in various places, civilizations embraced their own methods for rationalizing quantities and order. Our mathematical inheritance, including the language of mathematics, bears witness to this debt to the past – the base 60 of the Babylonians, the Roman system of numerals, the legacy of vigesimal systems all contribute to our vocabulary of numbers. It is a rich and multifaceted heritage.

WHY ARE PRIMES THE ATOMS OF MATHEMATICS?

Building blocks and the fundamental theorem of arithmetic

L ying within the counting numbers 1, 2, 3, 4, 5, etc. are the jewel-like prime numbers. They are building blocks of the whole number system, and the prospect of finding them and discovering their elusive properties has enticed generations of mathematicians from the earliest of times. Prime numbers beguile because they are so basic yet they suggest some of the toughest unsolved problems in mathematics.

So what are prime numbers? First we have the whole numbers 1, 2, 3, 4, 5, 6, etc., stretching out along the conceptual 'number line'. These are the counting numbers of mathematics, but among them lie the 'diamonds', so called prime numbers or 'primes'. Primes are the whole numbers that cannot be divided by any number exactly, apart from 1 and the number itself. The number 5 is a prime because dividing it by 2, 3 and 4 we always have a remainder left over. But 6 is not a prime because it can be divided by 2 to give an exact answer. That $6 = 2 \times 3$ means that it can be composed of smaller numbers, and mathematicians use the term *composite number* for a number that is not a prime. Composites can be broken down into smaller 'factors' but a prime like 5 is unbreakable. It is indivisible, as atoms were once supposed to be.

The first few primes are 2, 3, 5, 7, 11, 13, 17, 19, 23, 29, and so on, but prime numbers don't appear in a nice orderly fashion among the counting numbers. They are the mavericks of mathematics and appear to exhibit 'randomness' – though this has not prevented mathematicians searching for patterns among them and teasing out any regularity they might find.

Before we progress, there is a little legality to sort out. This concerns the number 1 itself. Is it a prime? Since it is not divisible by any number apart from itself it certainly conforms to the definition of a prime number. Indeed many famous mathematicians, including the great Leonhard Euler, have reckoned 1 as a prime number, though current thinking is to start prime numbers with the number 2. (If we accept 1 as a prime, some statements we would like to make would not apply to *all* prime numbers.)

Even if we accept the number 2 as the first prime, it too has special qualities. It is the only even number that is a prime: all even numbers are divisible by 2, so those even numbers greater than 2 cannot be prime. Apart from 2, therefore, when we look for primes we only need to consider odd numbers. Additionally, no prime with more than one digit can end in 0 or 5 because numbers ending in these digits can be divided by 5; so a prime greater than 10 must end in 1, 3, 7 or 9. But that doesn't mean that *all* numbers ending in these digits are primes: of the numbers between 1 and 100 that end in '1', the numbers 11, 31, 41, 61, 71 are primes, but 21, 51, 81, 91 are not.

The fundamental property of primes

There is a vast quantity of results known about prime numbers. 'Number theory', the oldest branch of mathematics, is packed with established theorems and future challenges, and each year hundreds of new theorems appear in the mathematical literature. The important results, the big ideas, illuminate the mathematical landscape on account of their simplicity and depth.

One crucial tenet is that *any* whole number can be expressed as the multiplication of primes. For example, we can represent the number 41,184 as a product of the primes $2 \times 2 \times 2 \times 2 \times 2 \times 3 \times 3 \times 11 \times 13$, which can be condensed, in the notation of 'powers', as $41,184 = 2^5 \times 3^2 \times 11^1 \times 13^1$. Thus, primes are the building blocks for the system of whole numbers. An allied property is that there is exactly *one way* of writing a whole number in this way: we could not have a set of different prime numbers multiplied together to give the same number. This uniqueness is called the fundamental theorem of arithmetic.

It follows from the fundamental theorem that any whole number can be divided by at least one prime number. There is at least one, but there are usually more, for example, the number 41,184, to return to our example, can be divided by the primes 2, 3, 11 and 13.

'We are bilaterally symmetrical, sexually differentiated bipeds located on one of the outer spirals of the Milky Way, capable of recognising the prime numbers …'

FROM TEXT ATTACHED TO NASA
DEEP SPACE PROBES

When is a number a prime?

It is easy to decide whether a *small* number is a prime number. Given a number N we can perform a few quick checks. And we can save a great deal of work by an observation. If a number N is a composite, that is, $N = a \times b$, then factors a and b cannot both be greater than the square root of N. If they were, then multiplying them together would give a number greater than N. So we may suppose one factor (say, a) is less than the square root of N. We also know by the fundamental theorem of arithmetic that a must be divisible by a prime. Putting these observations together, to see if N is a composite number we need only check whether any primes less than, or equal to, the square root of N actually divide into N. If none do then N must be a prime.

To investigate the number 211, for example, we must first calculate its square root, which is approximately 14.5. So we need only check the divisibility of 211 for primes less than 14, that is, check whether any of 2, 3, 5, 7, 11, 13 divide into 211 exactly. When we do this on the calculator and see that they all leave remainders, we conclude that 211 is a prime.

Obviously this method is impractical for numbers with thousands of digits, and checking whether a very large number is prime or composite is a challenging problem, for which there is little in the way of a simple test. In 2002, however, a major breakthrough occurred. It was proved by a trio of Indian mathematicians – Manindra Agrawal, Neeraj Kayal and Nitin Saxena – that there is an *efficient* algorithm, or recipe, for testing whether any number is a prime (see *Is There Anything Left to Solve?*).

Where do primes end?

The prime numbers are limitless. We may say there is an infinity of prime numbers, that is, whatever prime number is written down there is another one beyond it. The striking theorem that demonstrates this fact is stated in Euclid's *Elements* (Book 9, Proposition 20) as 'prime numbers are more than any assigned multitude of prime numbers'. The essential element in its proof is what is now termed a 'Euclid number'.

A Euclid number is obtained by multiplying together *all* the prime numbers up to a given prime (which we call P) and then adding 1. In mathematical terms, a Euclid number N has the form $N = (2 \times 3 \times 5 \times 7 \times \ldots \times P) + 1$.

When we come to $P = 13$, we find the Euclid number N is a non-prime, a composite number:

$$N = (2 \times 3 \times 5 \times 7 \times 11 \times 13) + 1 = 30,031 = 59 \times 509$$

In fact, prime Euclid numbers are very rare and we have to wait until $P = 31$ for the occurrence of another, producing the prime $N = 2,000,560,490,131$. After that we wait until we come to $P = 379$, and then the prime N would have approximately 200 digits. And after $P = 379$ the next prime Euclid numbers occur for $P = 1019$ and $P = 1021$, a pair of adjacent primes (called 'twin primes'). Writing down these gigantic Euclid numbers is out of the question.

Prime P	Expression	Euclid number N	Prime?
2	2 + 1	3	✓
3	(2 × 3) + 1	7	✓
5	(2 × 3 × 5) + 1	31	✓
7	(2 × 3 × 5 × 7) + 1	211	✓
11	(2 × 3 × 5 × 7 × 11) + 1	2,311	✓

EUCLID NUMBERS FOR SMALL VALUES OF P

We can now appreciate the genius of Euclid's claim that the number of primes is infinite. A proof goes like this: Let's choose *any* prime P. Our object is to prove there is a prime number beyond P. As we have seen, the Euclid number $N = (2 \times 3 \times 5 \times 7 \times \cdots \times P) + 1$ is either prime or not – but we don't know which it is for an unspecified value of P. If N is prime, we have our prime number beyond P, for certainly N is greater than P. On the other hand, if it is not prime we are not lost. On dividing N by any of the primes 2, 3, 5, 7, …, P, we always get a remainder of 1, so none of these divide into N. But by the fundamental theorem of arithmetic we know N must be divisible by some prime number, so since this guaranteed prime is not any of 2, 3, 5, 7, …, P, it must be greater than P. This will be the prime we are seeking beyond P.

We have thus proved that for any prime P there is always another beyond it – and the number of primes must therefore be infinite. If the prime numbers had been finite in number, we would have a totally different, and much poorer, branch of mathematics.

Special primes

Within the infinite set of prime numbers are some with a fixed form. Central to the lore of prime numbers are the Mersenne numbers, named after the 16th-century mathematician Marin Mersenne.

Mersenne numbers are expressed (using the notation n = any number) in the form $M_n = 2^n - 1$. They are always odd numbers because 1 is subtracted from the even number 2^n (which is the number 2 multiplied by itself n times). While they are odd, they are not necessarily prime. For M_n to be prime it is first necessary for n also to be prime – but this is not sufficient to guarantee that M_n is a prime. Although $M_2 = 3$, $M_3 = 7$, $M_5 = 31$, $M_7 = 127$ are all primes, the Mersenne number M_{11} is not a prime: it was discovered in the 16th century that:

$$M_{11} = 2^{11} - 1 = 2047 = 23 \times 89.$$

There seems to be no general pattern with Mersenne numbers. M_{13}, M_{17}, M_{19} are all prime, but M_{23} is not, because:

$$M_{23} = 2^{23} - 1 = 8{,}388{,}607 = 47 \times 178{,}481$$

In fact, most Mersenne numbers are not prime, and to date only 47 prime Mersenne numbers have been found. Finding out that $M_{11,213}$ was a prime was a source of pride for the University of Illinois in 1963. Since that time, much larger ones have been discovered, including the number $M_{43,112,609}$, a prime that weighs in at almost 13 million digits!

In traditional mathematics, prime Mersenne numbers can be used to construct *perfect numbers*, which are numbers that are the sum of their divisors. The number 6 is perfect because $6 = 1 + 2 + 3$, and 28 is the next one, because $28 = 1 + 2 + 4 + 7 + 14$. By combining the work of Leonhard Euler and Euclid, mathematicians have arrived at a formula for locating perfect numbers using prime numbers (p), so $2^{p-1} \times M_p$ produces a perfect number *provided M_p is a prime number.*

The frequency of primes

What, though, is the distribution of primes in any given interval? As we progress through the whole numbers, the primes become more sparse. A quarter of the numbers between 1 and 100 are primes, but between 1 and 1000 only 16% of numbers are primes, and between 1 and 10,000 only 12%. It has recently been verified that there are 1,925,320,391,606,803,968,923 prime numbers in the range from 1 to $n = 10^{23}$, which means that only 1.9% of the totality of numbers in this massive interval are actually primes.

As a boy of 16, the great early 19th-century mathematician Carl Friedrich Gauss was fascinated by tables of primes, and he set to work charting the proportion of primes in various intervals. This proportion, the so-called 'density' of primes, is worked out by calculating:

$$\frac{Total\ number\ of\ primes\ less\ than\ n}{Total\ of\ numbers\ less\ than\ n}$$

From his painstaking counting, he speculated on a theoretical *estimate* in terms of a formula.

Number (n)	Actual number of primes less than n	Actual density	Theoretical estimate
10	4	40.0%	43.4%
100	25	25.0%	21.7%
1000	168	16.8%	14.5%
10,000	1229	12.3%	10.9%
100,000	9,592	9.6%	8.7%
1,000,000	78,498	7.8%	7.3%
10,000,000	664,579	6.6%	6.2%
100,000,000	5,761,455	5.8%	5.4%
1,000,000,000	50,847,534	5.1%	4.8%

ACTUAL AND THEORETICAL DENSITY OF PRIMES

Gauss's formula (written in terms of the logarithm of n) was based on experimental evidence, but a mathematical proof of the formula was nowhere in sight at the beginning of the 19th century. By the end of the century, a proof for the theoretical estimate emerged, but it left some dissatisfied because it relied on Differential and Integral Calculus (see *What Is the Mathematics of the Universe?*), and the introduction of such techniques into pure 'arithmetic' was thought artificial. This criticism vanished 50 years later, when a proof by pure arithmetic was produced – though the argument is extremely technical and lengthy.

The established theorem is now referred to as the prime number theorem (PNT), and it says that the difference between the density of the primes and the theoretical estimate can be made as small as we want to by taking n large enough. The PNT is regarded by mathematicians as one of the greatest theorems in their subject, especially since it is bracketed with one of their greatest heroes, Gauss. It was his student, Bernhard Riemann, and his famous paper of 1859 that showed mathematicians how the long-term distribution of prime numbers depended on the behaviour of the 'Riemann zeta function' and the 'Riemann hypothesis'.

The search for regularity

Thanks to Euclid, the infinity of the primes has afforded a limitless playground for mathematicians. All kinds of questions arise, such as 'Is there any regularity in the appearance of the primes?' or 'Can we produce formulae that generate only primes?'

In the 18th century, Leonhard Euler produced a famous formula, $n^2 + n + 41$ (in which n, as ever, stands for 'any number'). It was special, because if we substitute $n = 1, 2, 3, \ldots, 39$ into the formula the answer is a prime number. If we choose $n = 7$, say, to test this out we get $7^2 + 7 + 41 = 97$, which surely is prime. In fact all goes well until we reach $n = 40$, when the answer is 41×41, a non-prime.

Expressions with many variables have achieved greater success, and it has recently been shown that there is a formula, albeit complicated, that will generate *all* the primes. The trouble is that it has ten variables and the powers attached to some of the variables can be very large (of the order 10^{45}). It will also produce numbers that are not prime.

Another type of success has been achieved through simply stated expressions of a different type. One spectacular one that is thought to generate only primes – but not *all* the primes – was discovered in 1947 by William H. Mills. He proved there is a number, designated k, such that:

The integer immediately below k^{3^n} is always a prime for any value of n.

The only difficulty is that no one knows much about the value of k, only that it exists, and there may be more than one value of it. It is now called the Mills constant. The smallest possible value of k is thought to be approximately 1.3063778838, if the Riemann hypothesis is true. Results on the distribution of primes we would dearly like to have depend on the truth of the Riemann hypothesis. Indeed, so much regarding prime numbers, and much else, depends on the Riemann hypothesis that its proof is eagerly awaited; it remains perhaps *the* most challenging problem in all of mathematics (see *Is There Anything Left to Solve?*).

The primacy of primes

We have, in our discussion of prime numbers, entered the waters of number theory, which has been claimed as the one branch of mathematics devoid of any applications. That might once have been the case, but now number theory is at the centre of the modern sciences of coding and computer security (see *Can We Create an Unbreakable Code?*). The security of computers depends on this recondite subject, where the theorems of Euclid, Mersenne, Euler, the 17th-century lawyer Pierre de Fermat – and a host of other theorists – have found a new lease of life.

The prime numbers are simply defined, but though much is known about them they continue to challenge in perplexing and tantalizing ways. They have attracted the attention of the world's greatest mathematicians, and it is undeniable that they are the building blocks of arithmetic and lie at the heart of so many mathematical endeavours. It would not be exaggeration to call them the 'atoms' of mathematics, and tracking down their remaining mysteries remains a fundamental quest of mathematics.

WHICH ARE THE STRANGEST NUMBERS?

Real, irrational and transcendental numbers

*E*arly civilizations laid the basis for our systems of counting, and the sequence of whole numbers beginning 1, 2, 3, in whatever system of symbols was adopted, proved an invaluable practical tool. Over time, though, this was insufficient to supply our mathematical needs. New types of numbers were needed, heralding the concept of numbers less than whole numbers – fractions – and numbers less than zero, that is, negative numbers. The elaboration of the number sequence did not stop there, producing some very mysterious numbers indeed.

The counting numbers that young children learn are the *whole* numbers (1, 2, 3, etc.). To remind themselves that they are the starting point for most investigations, mathematicians tend to call these the 'natural' numbers; they also refer to them as the *positive integers*, 'integer' meaning a whole number. (When zero is accommodated, the system is referred to as the *non-negative integers*.) Conceptually, it is useful to think of these basic building blocks of mathematics as strung out along a 'number line'.

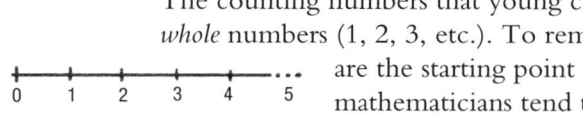

ZERO AND POSITIVE WHOLE NUMBERS
ON THE NUMBER LINE

But this is, in itself, inadequate for many mathematical calculations. The number line needs to be augmented with, for a start, *negative* numbers: −1, −2, −3, −4, −5, etc. One of the clearest ways we encounter negative numbers is in our financial affairs, when, depressingly, a figure of (say) '−250' on a bank statement indicates we are running an overdraft and owe the bank this amount. Negative numbers may also be used to measure temperatures below zero. The higher the negative number then the colder it is, so that −20 degrees is chillier than −10 degrees. This is a straightforward ordering of the negative numbers, and can be written as −20 < −10. When introduced into the number line, the negative numbers (negative integers) proceed from right to left as the actual number increases. The combined set of 'positive' and 'negative' integers produces what mathematicians call the *plain integers.*

PLAIN INTEGERS (NEGATIVE NUMBERS, ZERO, AND POSITIVE NUMBERS) ON THE NUMBER LINE

We now accept the existence of negative numbers quite happily, though in the past they had opponents who challenged their very existence. In the 18th century one Francis Maseres, an English diplomat who dabbled in mathematics, regarded negative numbers as nonsensical and meaningless. He declared that they 'darken the very whole doctrines of the equations'. It was a voice in the wilderness, however, and subsequent generations saw through his specious arguments to fix negative numbers firmly into the mathematical fabric.

From whole numbers to fractions

But negative integers could not solve the problem of how to represent parts of a whole number, which is where fractions come in. We often think of fractions as portions of a single entity. If 20 people attend a birthday party and there is one birthday cake, then each guest will, in theory, be entitled to one-twentieth, or $\frac{1}{20}$, of the cake. The top number is the *numerator*; the bottom number is the *denominator*. Fractions are often referred to

> *'Integers are the fountainhead of all mathematics.'*
>
> HERMANN MINKOWSKI,
> Diophantine Approximations (1907)

by mathematicians as *rational* numbers – not because they are inherently commonsensical, but because they are expressed as one number divided by another, that is, they represent a ratio.

Fractions of this kind, where '1' appears over any whole number (which we can describe algebraically as '*n*') are called *unit* fractions. The Ancient Egyptians developed interesting ways of calculating with them, prompting mathematical conundrums that remain unsolved to this day. With ordinary fractions, however, we don't restrict the top number to 1. So, if the birthday partygoers were presented with two identical cakes, they would each get $\frac{2}{20}$ of cake. Sometimes these fractions can be simplified by division, so that $\frac{2}{20}$, divided by 2, produces an equivalent fraction $\frac{1}{10}$ for the quantity of each birthday cake on offer to a single guest. This type of 'cancelling out' within fractions is a useful device and part of the art of handling fractions.

Returning to the visual device of the number line, we are able to represent fractions on it too: they can simply be inserted at the appropriate places between whole numbers, so that $\frac{1}{2}$ falls midway between 0 and 1, and a top-heavy fraction such as $\frac{5}{2}$ would fall midway between 2 and 3.

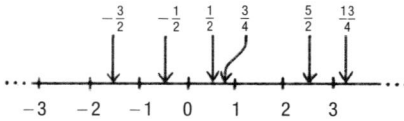

FRACTIONS ON THE NUMBER LINE

The number line raises a question though. Can we go on adding fractions into it? How many fractions can be squeezed in? There is actually *always* more space, so, for example, if we already had $\frac{2}{7}$ and $\frac{5}{17}$ on the number line, the fraction $\frac{7}{24}$ (formed by adding separately the denominators and numerators of the two fractions together) would lie between them. We might be tempted to say that fractions fill out the entire number line and there are no gaps left over.

Fractions and the Pythagoreans

Around 550 BC, the Greek scholars under the leadership of Pythagoras saw numbers as *the* key to nature. In one respect theirs was a quasi-mystical response; numbers were invested with human qualities, the even numbers regarded as feminine and the odd numbers as masculine. But the Greeks also had a hard-edged method of treating them, as was apparent in their theory of measurement, which focused on fractions.

In Pythagorean measurement, the length to be measured is divided by a given, smaller unit. The unit might be an inch, a millimetre or something else – the *name* of the unit is of no consequence for the theory. The unit is placed at the start of the length, and continually placed end to end – rather like measuring a distance by pacing it out, placing one foot directly in front of the other, toe to heel. If one were very lucky, a whole number of units, say 14, would make up the length. But, more likely, the final unit placed down would not fit flush with the end of the length – in which case the Pythagorean geometer would add *another* length, the same as the first (and so on), measuring until he did achieve a 'flush' fit.

He would then need to do some division to get the measurement of just one length. So if, say, 27 units fitted flush at the end of 2 lengths, the measurement of a single length would be the fraction $\frac{27}{2}$, which is $13\frac{1}{2}$ units. The curiosity of Pythagorean measurement was thus that it always produced a result *expressed as* a fraction.

The case of the vanishing fraction

In mathematics, Pythagoras's durable theorem for triangles is his chief claim to popular fame: that the square on the hypotenuse (the side in a right-angled triangle *opposite* the right angle) is equal in area to the sum of the squares on the other two sides. It is a theory that yielded implications for the practice of measurement (and fractions), and the Ancient Greeks thought in terms of literal squares, geometry, and not just a number multiplied by itself. Thus, Pythagoras's theorem was represented

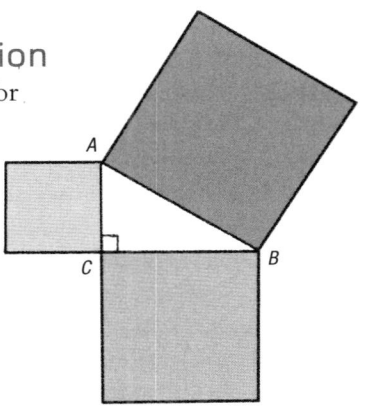

THE SQUARES OF A RIGHT-ANGLED TRIANGLE

39

by a diagram with real geometrical squares attached to the sides. If we denote the lengths of the sides by a, b, c, the theorem is better known by the equation $a^2 + b^2 = c^2$.

PYTHAGORAS'S THEOREM FOR
A RIGHT–ANGLED TRIANGLE

Let's imagine, though, that the length of both a and b is 1, which means that $1^2 + 1^2 = c^2$, so that $c^2 = 2$. The length of the hypotenuse must therefore be the square root of 2 (written as $c = \sqrt{2}$), which is approximately 1.4142 because 1.4142 × 1.4142 is 1.99996164. That is near 2, but is *not* exactly 2.

Now comes the shock. With Pythagorean measuring conventions, if we attempt to measure the side c (which should equal the square root of 2) using the side of the triangle $a = 1$ as our unit to place end on end, we find that no matter how many we place along the length c we will *never* get the unit finishing flush. This means that the side of a length defined as $\sqrt{2}$ cannot be a fraction. According to the prevailing theory of measurement, all lengths were supposed to be a fraction. But $\sqrt{2}$ evidently did not conform.

The Pythagoreans could not cope with this revelation. By legend, one of their number, Hippasus, was supposed to have been drowned at sea for divulging the secret of $\sqrt{2}$. Instead of the gods rewarding him, they vented their anger against the whistle-blower. But even the inhabitants of Mount Olympus could not alter the awkward reality about $\sqrt{2}$. Here was a length that could actually be constructed – what could be simpler than a right-angled triangle with two equal sides – yet its third side could not, by existing theory, be measured. Pythagoras might have been tempted to shrug off the inconvenience by saying that we should just persevere and lay down more units to achieve a result as a fraction. But his theory was demolished by the emergence of a logical proof that $\sqrt{2}$ was not a fraction and hence not measurable. There was therefore no way Pythagoras could dismiss this disclosure.

There were more unwelcome revelations ahead. After the square root of 2 failed to resolve into a fraction, a whole sequence of other square-root numbers presented a similar intractability: $\sqrt{3}, \sqrt{5}, \sqrt{6}, \sqrt{7}, \sqrt{8}, \sqrt{10}$, and so forth – missing out only the square roots of perfect squares like 4, 9, 16, 25, etc. We therefore find that, in addition to whole numbers and fractions (the rational numbers), we have inserted into the number line an infinite quantity of what mathematicians call *irrational* numbers, lurking between the fractions.

The rational numbers augmented with the irrational numbers make up what mathematicians call the 'real number line'. Any number can be located somewhere on it. But to define things negatively – to say that an irrational number is one that is *not* rational – does not reveal much about its nature. Richard Dedekind and George Cantor addressed this issue in the second half of the 19th century, and both came up with useful definitions that have a lasting place in mathematics. Dedekind defined a real number as a 'Dedekind cut', a 'cut' of the number line thereby identifying a real number as one which separates the number line into two parts; and Cantor defined real numbers as the limits of sequences of rational numbers, for instance $\sqrt{2}$ is the limit of the rational number sequence 1, 1.4, 1.41, 1.414, 1.4142, ... Through their work mathematicians are able to handle questions about real numbers with more precision.

From the irrational to the transcendental

The proof that mathematicians devised to determine the irrationality of the square root of 2 (and thus the square roots of other numbers) was an elegant one. But it produced little further insight into the *nature* of irrational numbers.

Cutting across the rational/irrational terminology is a different categorization, that of *algebraic numbers*. An *algebraic number* is one that can be a solution of an equation involving powers of x, such as $x^2 - 4x + 1 = 0$.

The irrationals $\sqrt{2}, \sqrt{3}, \sqrt{5}, \sqrt{6}, \sqrt{7}, \sqrt{8}, \sqrt{10}$, etc. are algebraic numbers, because, for example, $\sqrt{2}$ is a solution of the equation $x^2 - 2 = 0$ (and $\sqrt{3}$ satisfies the equation $x^2 - 3 = 0$, and so on). The category of algebraic numbers *includes* the categories of rational numbers and some irrational numbers.

Are *all* numbers algebraic numbers? The answer is 'no', for beyond the shores of algebraic numbers lie the 'wild' irrational numbers. These numbers do not satisfy *any* algebraic equation, and as such they were described in the 18th century as 'transcendental' by the Swiss genius Leonhard Euler. Quite simply, they 'transcended algebra'. But *showing* that a number is transcendental – i.e. that it is not the solution of *any* equation involving powers of x – is a distinct challenge, if only because proving a negative is notoriously difficult.

The category of transcendental numbers includes two of the most renowned numbers in mathematics. The most famous is 'pi', the number symbolized by the Greek letter π, the length of the circumference of a circle divided by its diameter, delivering a value (to the first six of its infinite decimal places) of 3.141592.

After π, the constant e ranks next in the transcendental hall of fame. This number, the infinite value of which (to six decimal places) is 2.718281, is occasionally known as 'Euler's number', for the mathematician who popularized it. It has since proved invaluable in such problems as working out population growth, in dealing with financial mathematics, and it occurs throughout probability theory and statistics. It was Charles Hermite, in the 19th century, who proved that e is transcendental. Modifying Hermite's technique, Ferdinand von Lindemann was then able to solve the 'riddle of the ages' by proving that π was transcendental too. With this discovery came confirmation of a famous conundrum: that it was indeed impossible to 'square a circle', that is, to construct a square with the same area as a given a circle using only a straight–edge and compass.

Around the time that Lindemann was making his conclusive investigations of π, Georg Cantor produced something of a mathematical sensation. Mathematicians were familiar with fractions and irrational numbers, and a few transcendental numbers like e and π. Cantor proved that almost all 'real' numbers – numbers which are either rational or irrational – are transcendental, prompting a rethink on the nature of the traditional number line. If we thought the number line consisted of only the fractions and whole numbers, plus numbers like $\sqrt{2}, \sqrt{3}, \sqrt{5}, \sqrt{6}, \sqrt{7}, \sqrt{8}, \sqrt{10}$, etc., we could not be more wrong. These are in fact quite rare when compared with the transcendentals clustered around them.

With the revelation that transcendental numbers are far more numerous than any other kind of number it remains curious that, apart from π and e, we can't actually specify many of them. We can play games and manufacture some if we wish, like 7π, $\pi^2 + 1$, e^2, $e^3 + 1$, but this is scraping away at the surface. We have to be careful too. If we start mixing transcendentals arbitrarily we could run into difficulty. It is not even known, for example, whether the number $e + \pi$ is transcendental or not. The number e raised to the power π is transcendental, but it is not known whether the same could be said about π raised to the power e.

Nevertheless, there has been progress. In 1900 the mathematician David Hilbert set his colleagues a famous challenge of '23 Problems' for the new century. The Seventh Problem was to show that $2^{\sqrt{2}}$, that is, 2 to the power of the square root of 2, is transcendental. Thirty years later the problem was solved by Aleksandr Gelfond, who also discovered whole families of transcendental numbers.

The impact of decimals

Mathematicians tend to use letters, like e and π, and square root signs for some of their numbers, but for practical purposes decimals are needed. At its simplest, a quarter $\left(\frac{1}{4}\right)$ can be expressed as 0.25, or $\frac{3}{5}$ as 0.6. But of course not that many fractions can be converted so neatly. One-third $\left(\frac{1}{3}\right)$, represented decimally, produces an infinitely receding 0.33333… However, what we can say about the decimal expansion of a fraction is that it has a regularity, a *recurring* pattern. The fraction $\frac{2}{7}$, for example, represented decimally becomes 0.285714285714285... It repeats the sequence 285714 forever.

Decimals that represent irrational numbers are also infinite, but are *patternless*, though this distinction does not work to distinguish between algebraic numbers and transcendentals. The algebraic number $\sqrt{2}$ and the transcendental π both have patternless decimal expansions:

$$\sqrt{2} = 1.414213562373095...$$

$$\pi = 3.141592653589793...$$

Continued fractions as DNA

Looking at the *continued fraction* expansions of numbers is deeper. It can be regarded as a geometrical theory. Let's work out the continued fraction expansion of $5\frac{29}{94}$. Focusing our attention on the $\frac{29}{94}$ part we might imagine a rectangle of short side 29 and long side 94.

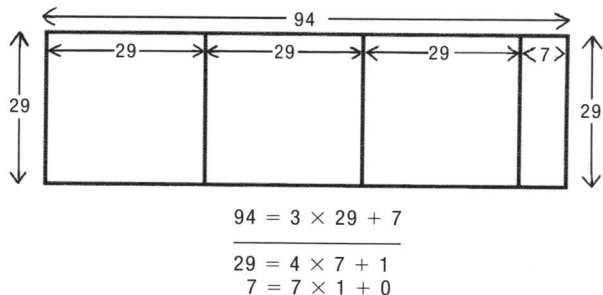

$$94 = 3 \times 29 + 7$$
$$\overline{29 = 4 \times 7 + 1}$$
$$7 = 7 \times 1 + 0$$

THE CONTINUED FRACTION EXPANSION OF NUMBERS

We are able to subtract a maximum of three squares (of side 29) from this rectangle leaving as a remainder a rectangle of short side 7 and long side 29. If we repeat the operation on the remainder we find we can subtract four squares (since $29 = 4 \times 7 + 1$) and finally seven unit squares. The continued fraction of $5\frac{29}{94}$ is determined by the number of squares we can subtract (3, 4, 7), and is presented as:

$$[5; 3, 4, 7]$$

These continued fractions are like the DNA of a number. Under this microscope, $\sqrt{2}$ has a *regular* continued fraction representation:

$$\sqrt{2} = [1; 2, 2, 2, 2, 2, 2, 2, 2, \ldots]$$

This is because we can subtract two squares at each stage and never come to an end. It is very regular while the transcendental e has a continued fraction expansion also showing signs of a different sort of pattern:

$$e = [2; 1, 2, 1, 1, 4, 1, 1, 6, 1, 1, 8, 1, 1, 10, 1, 1, 12, 1, \ldots]$$

The celebrated π remains an enigma. It has a highly irregular continued fraction expansion and this wild number shows no sign of settling down:

$$\pi = [3; 7, 15, 1, 292, 1, 1, 1, 2, 1, \ldots]$$

The expanding horizons of numbers

When human beings wished simply to quantify physical entities before them – their livestock, their crops, their lands – there was a sense in which the ordinary counting numbers were sufficient. But the handy conceptual tool of the number line has undergone fundamental transformations over the centuries. The emergence of negative numbers meant that the number line not only increased, from left to right, but decreased from zero, right to left. The mathematical dilemma that the square root of 2 posed to the Greeks of the Pythagorean school effectively shattered a comforting mathematical mindset of fractions, the mathematician's rational numbers. The world of numbers took on strange and uncertain forms, expanding into what are today called the 'irrational' numbers, and the much-theorized but rarely specified transcendental numbers. It is all a long way from tally sticks.

ARE IMAGINARY NUMBERS TRULY IMAGINARY?

From the imaginary 'i' to octonions

*T*he Romantic poet Samuel Taylor Coleridge, who espoused the powers of human imagination and personal experience, was circumspect about mathematics, a discipline in which he felt 'reason is feasted, imagination is starved'. But two centuries earlier, the philosopher Descartes was already speaking of a mathematical realm of 'imaginary' numbers. Treated at first with extreme caution, 'imaginary' numbers were eventually accepted as part of the mathematical fabric.

The methods that Newton and Leibniz had devised for calculating rates of change in the Calculus took on a new form, and mathematicians created what they called 'complex analysis' as a powerful technique in geometry and the theory of numbers. The novel systems of algebra discovered as extensions of the 'imaginary' challenged preconceptions about the nature of algebra itself. And perhaps the most unlikely development was the sheer usefulness of 'imaginary' numbers in real applications.

Where, then, are these imaginary numbers? Some are more familiar than we might think. We can take a basic scenario to illustrate the idea at its simplest, using the straightforward counting numbers. If we have 5 apples in a fruitbowl and we take away 4, there is 1 left. If we take away all 5, however, we have 0 apples left over. But we cannot take away 6 apples because there are only 5 apples in the fruitbowl to start with.

In our banking transactions we can become worryingly familiar with a different kind of notional number – the amount we owe. Here, we might have £5 in our account, but we also have an overdraft facility. So we withdraw £7 and owe the bank £2, which might show up on a statement as −£2. We cannot actually see or hold the negative amount −£2 of course – in that sense, it does not exist – but we certainly know how to interpret our debt, and so does the bank. And that means we can efficiently calculate with an imaginary −£2 coin, so that if we pay in £10 the next day we know that £2 will be taken by the bank to compensate for its loan and our new balance will be £8. In mathematical notation, the sum is $10 + (-2) = 8$, which is the same as saying $10 - 2 = 8$.

Looked at this way, the negative numbers could be considered 'imaginary', and when they first came about they were so regarded, as we saw earlier (see *Which Are the Strangest Numbers?*). Now they are uncontroversial, and by admitting them to mathematics we helpfully enlarge the number system. In mathematical theory, we can conceptualize negative numbers and their usefulness in terms of equations. If we were limited to the positives 0, 1, 2, 3, etc., we could solve equations such as $x - 1 = 0$. The solution would be $x = 1$. But we would not be able to solve $x + 1 = 0$. If we substitute any of the numbers 0, 1, 2, 3, etc. for x we will fail to find a result equalling zero. We have to go *outside* the positive numbers for a solution, for it is only when we admit negative numbers that we get the solution, which is $x = -1$. By imagining the concept of a negative number, we have, in a mathematical sense, brought it into existence.

Quadratic equations and the imaginary 'i'

Imaginary numbers took a step further in the context of the 'quadratic equation', a phrase that lingers in the memories of many an adult from their school lessons. Quadratic equations, an old type of problem, were solved by the Ancient Egyptians as they calculated the areas of fields. The defining quality of a quadratic equation is that it involves an unknown number that is squared, represented as x^2, so that from the idea of a four-sided square came the description 'quadratic'. Applied to real, actual phenomena, the mathematics involved in quadratic equations pose no particular problems.

But what happens in a quadratic equation such as $x^2 + 1 = 0$? For the equation to work in an ordinary way, we require x^2 to be the negative number -1 to produce a result of 0. But that cannot happen, because x^2 can never *be* a negative number. (If say, $x = 5$, then $x^2 = 25$; and if $x = -5$, then x^2 is also 25!)

A further leap of the imagination is needed to overcome the problem. And it lies in *inventing* a solution to $x^2 + 1 = 0$ and defining x by a name, which mathematicians have called 'i', standing for imaginary. As we have declared it a solution, so $i^2 + 1 = 0$, or in another words $i^2 = -1$. The word 'imaginary' was first used in this mathematical context by Descartes in the 17th century. He thought that all quadratic equations *should* have solutions, but accepted that 'in many cases no quantity exists which corresponds to what one imagines'.

All we have claimed about the symbol i is that $i^2 = -1$ and that is all we need to know. We need spend no time on questions surrounding the nature of i. It would thus be a mistake to take this one stage further — as some have done — and describe i as the square root of -1. For that would be to imply that it has existence as a real number along the number line, as if we could plug -1 into a calculator, press the $\sqrt{}$ button and obtain a real-number answer.

The symbol allows us to produce solutions to an array of quadratic equations that have no real-value solutions. The equation $(x - 1)^2 = -4$, for example, can be written out as:

$$\left(\frac{x-1}{2}\right)^2 = -1$$

That enables us to identify the expression inside the bracket with the imaginary i, which gives a solution $x = 1 + 2i$. In fact, this equation has two solutions, the other being $x = 1 + (-2)i$.

Complex numbers

Mathematicians call the solution of the type we have just achieved, $x = 1 + 2i$, a *complex number*. To generalize, the formulation $a + bi$, where a and b stand for real numbers, is 'complex' in that sense. But one could be forgiven, at this stage, for wondering whether we have entered a realm of fantastical solutions invented to serve fanciful problems. The air of mystery surrounding complex numbers was partially dispelled when it was realized that they possessed concrete meaning in geometry. Important to this development was the work of the Irish prodigy William Rowan Hamilton, who, writing in the 1830s, advanced the idea of treating the complex number $a + bi$ as a couple (a, b), stripping away any mention of i.

So mathematicians have two ways of writing a complex number. When they want to justify complex numbers on Sundays they could write (a, b) but during the working week write them as $a + bi$ and deal with them accordingly.

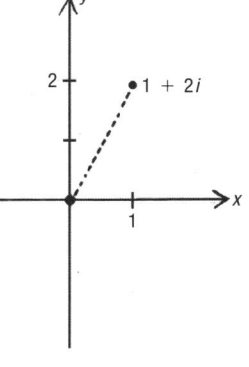

Real numbers, as we have seen (see *Why Are Primes the Atoms of Mathematics?*), lie on the one-dimensional number line, moving right (the positive numbers) and left (the negative numbers). But complex numbers are two-dimensional. A complex number can be represented by a point in a plane, in which case the resulting diagram is called an Argand diagram, after one of its early 19th-century progenitors, the amateur French mathematician Jean-Robert Argand.

THE COMPLEX NUMBER $1 + 2i$, A TWO-DIMENSIONAL POINT

Using the example of the complex number $1 + 2i$, we can show the complex number in a graphic representation, as the point with x coordinate equal to 1 and y coordinate equal to 2. In the conventions for writing coordinates, the point would simply be known as $(1, 2)$.

That is not the end of the geometric usefulness of the *i* though, for it has further meaning. If we multiply any complex number by *i* we obtain another complex number at right angles to the original one. For instance, using our exemplary number of $1 + 2i$:

$$i \times (1 + 2i) = i + 2i^2 = i - 2 = -2 + i$$

The effect of multiplying by *i* is therefore to *rotate* the point $(1, 2)$ in the plane into the point $(-2, 1)$; see illustration.

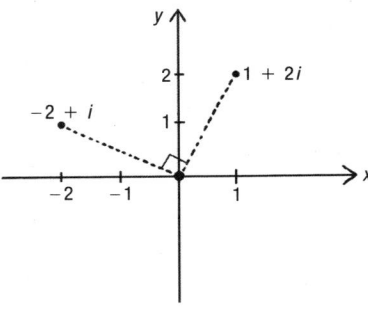

THE COMPLEX NUMBER $-2 + i$ IS A ROTATION OF $1 + 2i$

The symbol *i* thus generates *two* geometric meanings, as a *point* in the plane or as a *rotation* through a right angle.

Just as with numbers on the number line, complex numbers can be added, subtracted, multiplied and divided. They represent a further enrichment of our number system. A whole new branch of mathematics – complex analysis – has emerged from the world of complex numbers, and it has turned out to be more elegant and far reaching than the comparable theory based on real numbers. It is a form of the Calculus (see *What Is the Mathematics of the Universe?*), with variables that are complex numbers, in distinction to the older Calculus in which the variables are real numbers. This newer theory makes many links with other parts of mathematics, including geometry, and in the 19th century Bernhard Riemann applied it in his theorizing on the distribution of prime numbers.

There have also emerged a myriad of applications for complex analysis. Aircraft design is just one of them. The outline of the Joukowski aerofoil is traced out if we conceive of *z* as a point in an Argand diagram which lies on a circle with radius 1 and pinpoint the values of:

$$z + \frac{1}{z}$$

Complex analysis sees its most notable, indispensable, real-world application in the work of electrical and electronic engineers – though they use the symbol j instead of i, to avoid confusion with i, a symbol they reserve to represent electrical current.

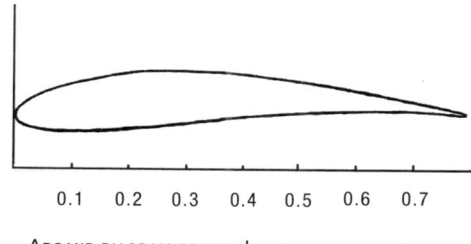

0.1   0.2   0.3   0.4   0.5   0.6   0.7

ARGAND DIAGRAM OF THE JOUKOWSKI AEROFOIL

From complex numbers to quaternions

Having contributed to the theory of complex numbers in a significant way, Hamilton then wondered whether it was possible to extend them, to broaden the repertoire. Mathematicians had widened the concept of numbers from counting numbers to negative numbers, and from real numbers to complex numbers. Could they go further?

The way ahead, as Hamilton saw it, was to move from the two-dimensional complex numbers to three-dimensional numbers. Years of thought produced nothing, but in 1843 he had a flash of inspiration. He had been on the wrong track and instead of searching in *three* dimensions he should have tried *four*. The result was Hamilton's 'quaternions'.

A quaternion is expressed in the form of the notation $a + bi + cj + dk$, and is analogous to a complex number $a + bi$. Like a complex number with $i^2 = -1$ the 'imaginaries' i, j, k of the quaternions have the properties $i^2 = -1, j^2 = -1, k^2 = -1$.

As an aid to multiplying quaternions, the symbols i, j and k may be conceived geometrically as points on the circumference of a circle, in which each letter multiplied by the next letter along equals the third letter. *But,* multiplying anti-clockwise produces positive values, whereas multiplying clockwise produces negative values, so that $i \times j = k$ (anti-clockwise), but $j \times i = -k$ (clockwise).

What many mathematicians found hard to accept was that the *order* of multiplication mattered. In ordinary arithmetic we get the same answer whether we multiply 7 by 5 or 5 by 7, and the same applies to the two-dimensional complex numbers. But the novelty of this situation did not perplex Hamilton, and it proved to be path-breaking. With quaternions, a new type of algebra was born.

There was still the question of how quaternions related to the question of rotation, a key property of the complex numbers. Shortly after the formal rules of manipulating quaternions were written down, Hamilton found that they could indeed be used to describe a rotation in three dimensions, and the way to do it was to identify a point (b, c, d) in three dimensions with the 'imaginary part' $bi + cj + dk$ of a quaternion.

From quaternions to octonions

Once the quaternions were established, the race was on to construct other systems. The real numbers on the number line are one-dimensional, the complex numbers two-dimensional and the quaternions four-dimensional. This progression of one, two, four – doubling each time – suggested the clue as to the next step.

A young English mathematician, the Cambridge-educated Arthur Cayley, had been attracted to quaternions and, in the wake of Hamilton's discovery, demonstrated how they could be used to define a rotation. He went on to frame a theory of eight-dimensional numbers, referred to as octonions or 'Cayley numbers'.

The octonions (w) are of the form $w = a + bi + cj + dk + ep + fq + gr + hs$, where the seven imaginaries (i, j, k, p, q, r, s) all have a square whose value is -1. The imaginaries for the octonions are multiplied according to the rules of a finite geometry called the Fano Plane, consisting of seven lines (the central circle is taken as a line) and seven points (see also *What Is Symmetry?*).

In this geometry, i, j and k, positioned around the circumference of the interior circle, resemble the geometry of the quaternions and are multiplied in the same way, taking note of the directions indicated by the arrows. If we want to calculate

$k \times p$ we look for the line containing k and p, note its direction and equate it to the third imaginary with the appropriate directional sign, which is s.

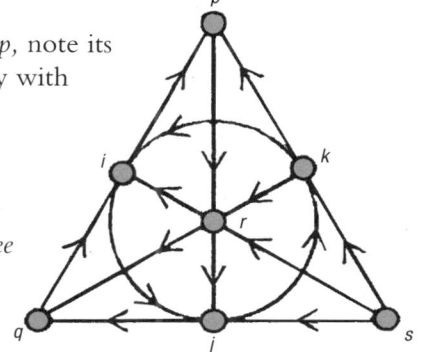

THE FANO PLANE FOR HANDLING OCTONIONS

With the quarternions the order of multiplication matters, but the revolutionary part of the octonion system occurs when *three* octonions are multiplied together. It matters which two are multiplied first!

The more adventurous mathematicians accepted these deviations from the usual rules of mathematics, and by the 1840s it was firmly established that there were imaginary systems in one, two, four and eight dimensions. Following the doubling principle, could this be pushed further – to a 16-dimensional system? The answer came many years later: there is no point in trying to construct an algebra in 16 dimensions. The real numbers, the complex numbers, the quaternions and the octonions turn out to be the only possible systems in which addition, subtraction, multiplication *and* division are possible.

The use of quaternions for rotations has since found applications in navigation, robotics and in the design of computer games – indeed, in many technological fields where the manipulation of rotations is required. But perhaps the real value of quaternions and octonions to mathematicians is to suggest a wider vista for other types of algebra.

Clifford algebras

If we drop the requirement that division should always be achievable, other possibilities arise. Clifford algebras, named after English 19th-century mathematician William K. Clifford, take the complex numbers and quaternions but go in a different direction from the one that led to the octonions. A Clifford algebra can also be described in terms of a number of 'imaginaries' whose square is −1. They are essential for physics and have found application in quantum theory. In Paul Dirac's 1928 paper 'The Quantum Theory of the Electron' his famous equation for the electron used these imaginaries to unlock the secrets of electron spin.

From imagination to application

What started off as distrust for the imaginary *i* of complex numbers has changed mathematics in such a palpable way that the mathematics in existence before its appearance seems uni-dimensional and poverty stricken. Descartes may have invoked the word 'imaginary', but the resulting beautiful construction of complex numbers and their subsequent generalizations have transformed mathematics, made a major impact on the study of physics, and have proved indispensable in new technological fields.

Once admitted into mathematics the imaginaries took on a life of their own. They challenged notions of algebra and what was permissible. Cross-links were found with the other developments in algebra during the 19th century, and the goldmine is still revealing its riches. But we have to be prepared to suspend what is obviously 'reasonable' and take leaps in the dark.

Imaginary numbers are not real if we expect them to lie on the one-dimensional number line containing the numbers we use in everyday life. But if we allow for two-dimensions, the imaginary and complex numbers are as real as any others. Mathematicians today could not function without them. And physicists have found in them the essential building blocks for their theories of the universe. In that sense, imaginary numbers have had profound consequences. Had Coleridge had the chance to look into a crystal ball, he might have seen that *because* imagination was feasted in the end reason was too.

HOW BIG IS INFINITY?

Set theory and the infinity revolution

*I*t is difficult to get our heads around infinity. The concept
appears vast and unfathomable, and consequently disconcerting.
How can there be an unending amount of anything – doesn't
everything, eventually, reach its limit? But if it does, what is
beyond it? Infinity has puzzled philosophers for centuries, and
in more recent times has fascinated physicists, astronomers and
cosmologists as they attempt to understand the universe. But
mathematicians have cast new light on the concept of infinity
and, more than that, with set theory they have raised the
prospect of a range of infinities.

'Infinity' is a term packed with mystery. Its close relative is
'eternity', time unending, and from an early age we wonder
about these scarcely graspable ideas. By tradition, for the
religiously minded, including many philosophers, there has been
a higher power to whom these unknowables may be entrusted.
In that sense, God *is* infinity and eternity – there is no time and
space outside Him.

But the requirements of mathematics are different, and
how do we account for infinity and define it *mathematically*? In
theory, infinity *ought* to be an easier concept than eternity, since
the time element is removed.

The unattainable infinite and worlds 'without bound'

If we were to ask a small child what is the biggest number they know, they might say something like 100, or, when they are a few years older and with more confidence, a billion billion. They are on the right track but we can always reply that 101 or a billion billion plus 1 are even bigger numbers. After a while they might be convinced that there is no biggest number and jump to 'infinity' as the answer to end all doubt. They are actually being quite revolutionary in this, by thinking that 'infinity' is an actual number.

This is not the conception of infinity held by the Ancient Greeks of the third century BC and the legions of mathematicians following them. They conceived of a 'potential infinity'. For Aristotle and others, infinity was unattainable, so it was *not* an actual number. In stating that there was an 'infinity of prime numbers', the geometer Euclid said that 'prime numbers are more than any assigned multitude of prime numbers', invoking the idea of an endless quantity.

We use the special symbol ∞ for this potential infinity, introduced by John Wallis in the 1650s. It could have been adapted from ω (omega), the last letter in the Greek alphabet, but its true origin is obscure. To emphasize that ∞ is not a number, mathematicians do not usually write $n = \infty$. Infinity is used in the sense of a number n becoming larger and larger, and they will say *n approaches* infinity and write it as $n \to \infty$.

Infinity, however unattainable, remained central to the mathematician's world. The big idea was the opening up of its meaning but this only came about through some revolutionary thinking.

A working definition of infinity, to get us off the mark, is of the kind we might find in a mathematics textbook, where infinity 'refers to a quantity without bound'. This conforms to the idea of 'bigger and bigger' because the whole numbers 1, 2, 3, ... are quantities that go on forever without bound.

A chosen number can always be superseded. But if the whole numbers are infinite, what about the fractions of the number 1:

$$\frac{1}{2}, \frac{2}{3}, \frac{3}{4}, \frac{4}{5}, \frac{5}{6}, \frac{6}{7}, \frac{7}{8}, \frac{8}{9}, \frac{9}{10}, \cdots$$

Surely there are as many of these as there are whole numbers? They are not 'without bound', because we know they are all less than 1. But it remains true that there is still an infinite number of them.

Set theory

We really need a more sophisticated definition of infinity than 'without bound' to open up the secrets of infinity, and that is where set theory comes in. This mathematical investigation was initiated by Georg Cantor in the 1870s. A *set* is intuitively a collection of things. So, for example, the positive whole numbers taken in their entirety is a set. To clarify this we use brackets { } and give the set a notation, which for the whole numbers is N, and write it:

$$N = \{1, 2, 3, 4, 5, 6, 7, 8, 9, \ldots\}$$

In this way we can think of the set as a single entity.

We can also talk about *subsets* of N. We could, for example, have the subset $A = \{1, 2, 3, 4, 5\}$, which we would call a *finite set*, because it is limited by its largest element, the number 5. This gives us a backdoor way of defining an infinite set as one that is *not* finite.

Relationships of sets

There is a striking property of N. Let's ask a naive question: are there as many *even* numbers, designated as the subset $E = \{2, 4, 6, 8, 10, 12, \ldots\}$, as there are positive whole numbers? The intuitive answer is no, because the even numbers are only a part of the totality of whole numbers. But what do we mean by 'as many', for surely we cannot *count* the number of elements in either set E or N and compare answers – they are both infinite sets!

This is where the notion of a one-to-one correspondence proves useful. To each whole number there 'corresponds' an even number, and vice-versa. How does this work? Each whole number corresponds to its double, so 1 in set *N* corresponds to 2 in set *E*, 2 corresponds to 4, etc., and every number thus achieves a correspondence.

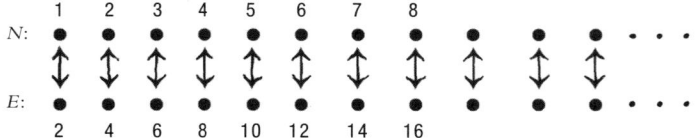

THE CORRESPONDENCE BETWEEN SETS *N* AND *E*

The other way around we can say each *even* number corresponds to half a whole number, so that 8 in set *E* corresponds to 4 in set *N*. Because this one-to-one correspondence exists for the two sets, they are described by mathematicians as *equinumerous*.

Galileo made the same sort of deduction about the squares of whole numbers, designated as set $G = \{1, 4, 9, 16, 25, 36, \ldots\}$, as there is a one-to-one correspondence between *G* and *N*. Some strikingly sparse sets are also equinumerous with *N*, for example the set *T*, where the whole numbers acquire an extra 0 as they progress up the number line:

$$T = \{1, 20, 300, 4000, 50{,}000, 600{,}000, 7{,}000{,}000,\\ 80{,}000{,}000, \ldots\}$$

In this subset of *N*, the distance between numbers is ever increasing, but a one-to-one correspondence between *N* and *T* remains clear, and these two sets are also equinumerous.

These examples nevertheless retain their paradoxical qualities. They appear to lead to counterintuitive statements and fly in the face of the common notion, as understood by the Ancient Greeks, that the 'whole is greater than the part'.

Checking in at Hilbert's Hotel

'Hilbert's Hotel', run by its patron the mathematician David Hilbert, was a thought experiment that exemplified this unusual behaviour of infinite sets. This notional hotel boasts an unlimited number of rooms. We suppose that, by 7p.m. one evening, the hotel is 'full', that is, for each number n, Room n is occupied. At 8p.m. a new guest arrives. What should Hilbert do? Should he show his new customer the 'Hotel Full' sign or try to find some extra accommodation? He puts on his mathematical hat, sees a way around the problem and starts shuffling his existing guests. He asks the guest in Room 1 to move to Room 2, the guest in Room 2 to move Room 3, and so on, each guest moving to the room next door. The new visitor is allocated Room 1, which is now spare. Problem solved!

At 10p.m. something more drastic happens. Coaches arrive bearing an unlimited number of people wanting accommodation. Hilbert is calmness itself. He moves his guests again, this time by doubling their room number. So the guest in Room 1 is moved to Room 2, the guest in Room 2 is moved to Room 4, and so it continues. After this reallocation the odd numbered rooms are now free and he happily allocates them to the newly arrived coachloads. Even if another unlimited number of guests arrive later on he can still accommodate them by repeating the same operation. There is no limit to accommodation in the Hilbert Hotel.

The patron David Hilbert is a very orderly man and he finds it very useful to have a list of all the people staying at the Hilbert Hotel. So, against the room number he writes the occupant's name. True, it will be an infinite list, but at least he will be able to tell who is the occupant in each of the rooms.

Making a list, checking it twice
The idea evoked by Hilbert, of constructing a list, has proved central to mathematicians' understanding of infinity. The set N can be looked on as a list 1, 2, 3, 4, etc. Subsets of N can be put in a list, but are there any *larger* sets that can be put in a list? The answer is yes – the set of whole numbers Z, which includes the *negative* whole numbers and positive whole numbers:

$$Z = \{\ldots -5, -4, -3, -2, -1, 0, 1, 2, 3, 4, 5, 6, \ldots\}$$

How can we put Z in a list? Along the number line, Z stretches from $-\infty$ to ∞ so how can we make it into a list? One way is to choose 0 as the beginning number, 1 as the second, -1 as the third, and so on, alternating descending and ascending numbers, so that our list is:

$$0, 1, -1, 2, -2, 3, -3, 4, -4, 5, -5, 6, \ldots$$

More impressively, we can even put *all* the fractions, designated as Q, into a list. It is not in numerical order but it is definitely a list. By way of experiment, we could try writing down the whole numbers (written with denominators 1) followed by the fractions with denominators 2, then those with denominators 3, and so on:

$$0, 1, -1, 2, -2, 3, -3, 4, -4, 5, \ldots$$

$$\frac{1}{2}, -\frac{1}{2}, \frac{3}{2}, -\frac{3}{2}, \frac{5}{2}, -\frac{5}{2}, \frac{7}{2}, \ldots \qquad \frac{1}{3}, -\frac{1}{3}, \frac{2}{3}, -\frac{2}{3}, \ldots$$

Is this a list? We can tell which is the first in the list, the second, third and fourth but what is the position of $\frac{1}{2}$? The trouble is that we can't hop across the ellipsis '...' between categories. But all is not lost. We can first put the fractions of Q into a *two-dimensional* array. On the top row will be the whole numbers; the next row will be the fractions with 2 as a denominator (omitting repeated numbers like $\frac{4}{2}$, which equals 2 and is already on the line above). In the third row we will put all the fractions with denominator 3, leaving out redundancies as we go.

We can now manufacture a list by starting in the top left-hand corner at 0, moving across (to 1) but then starting a diagonal movement to $\frac{1}{2}$ (see diagram). Doing the 'diagonal jig' enables us to write down our list as:

$$0, 1, \frac{1}{2}, \frac{1}{3}, -\frac{1}{2}, -1, 2, \frac{3}{2}, -\frac{1}{3}, \frac{1}{4}, \frac{1}{5}, -\frac{1}{4}, \frac{2}{3}, -\frac{3}{2}, -2, 3, \frac{5}{2}, -\frac{2}{3}, \frac{3}{4}, \ldots$$

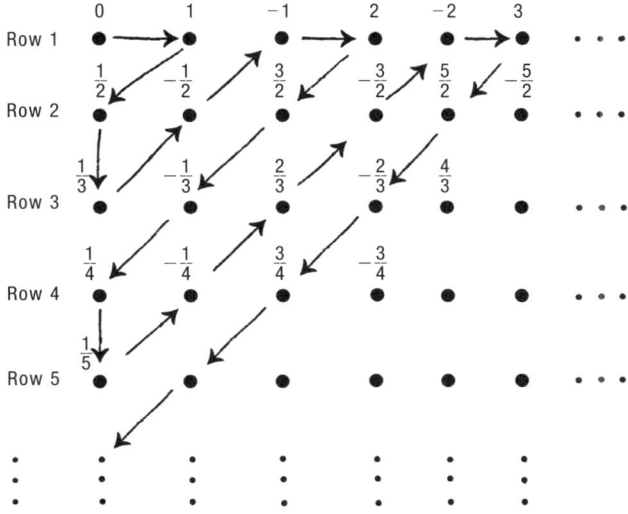

Each fraction is somewhere in the list and we can always tell where it is and identify its corresponding whole number. The list is not in numerical order, but it is a list.

Both N and Q are infinite sets and because they can both be written in lists they are equinumerous. Moreover, any set of numbers which can be put into a list is equinumerous with N.

A set without a list

What can be said about the set of what mathematicians call real numbers, the numbers R, which can be arranged along the number line? R includes numbers like the constant π and the other irrational numbers (see *Which Are the Strangest Numbers?*) and is called the real number *continuum* because the number line has no gaps in it. Like set Q, the continuum R is an infinite set; but can we arrange the numbers in R in a list? If we could, then R would be equinumerous with Q.

In the 1870s, Georg Cantor discovered one of the most astounding results in mathematics. In effect, he showed that the real numbers *cannot* be written down in a list. His argument goes like this. Every real number has a decimal expansion, so, for example, we might have the number, 34.8967494... If we want to *create* a number different from this we could ignore the whole number 34 and change the first number after the decimal point to 1, giving us 34.1967494... The fact that this number differs in just *one* decimal place from 34.8967494... is enough to make it a different number. (And if the first number after the decimal point had been 1 we could just substitute a 2.)

Suppose now there is a complete list of the real numbers *R*. Using the method we have described we can construct a number differing in the first digit from the first number in the list, differing in the second digit from the second number in the list, and so on, the result being that the number we have created cannot appear *anywhere* in the list because it is different from all of them.

The inescapable conclusion is that there cannot be a complete list of the real numbers *R*.

By adding the fractions into the set of whole numbers *N* we formed a larger set *Q*, but it was the one with the same *type* of infinity as *N*. But when we add the irrational numbers to the fractions, we get in the set *R* a *higher order of infinity*. This was revolutionary. There was not one infinity but two, the order of infinity of *Q* and the higher order of infinity of *R*. Cantor's theory went against the ideas of some well-established mathematicians.

Orders of infinity and the continuum hypothesis

Faced with different orders of infinity, Cantor took the idea of a *cardinal* number as a means to distinguish them. The cardinal number of a set is broadly its size, so, for the set $\{a, b, c, d, e\}$ the cardinal number is 5, because there are 5 elements. But what do we do about infinite sets? Cantor denoted the cardinal number of *N* by $\aleph_0$, where $\aleph$ (aleph) is the first letter of the Hebrew alphabet. The notation $\aleph_0$, vocalized 'aleph nought', has remained

> 'The fear of infinity is a form of myopia that
> destroys the possibility of seeing the actual infinite,
> even though it in its highest form has created and
> sustains us, and in its secondary transfinite forms
> occurs all around us and even inhabits our minds.'
>
> GEORG CANTOR,
> *Essay on Actual infinity* (1886)

unchanged since Cantor's day. The cardinal number of set R is denoted by c, from 'continuum', and because the infinity of N is of a lesser infinity than R we can write $\aleph_0 < c$.

Can we squeeze a cardinal number in between $\aleph_0$ and c? In ordinary arithmetic we have inequalities, for instance $\frac{1}{2} < \frac{4}{7}$, and indeed between any two fractions we can always squeeze another. A neat way of doing this is to add the numerators and denominators together to get $\frac{1+4}{2+7} = \frac{5}{9}$ and we find $\frac{1}{2} < \frac{5}{9} < \frac{4}{7}$. But if we just consider whole-number inequalities such as $2 < 3$, we cannot squeeze in another whole number.

A hypothesis – the continuum hypothesis – states that there is no cardinal number between $\aleph_0$ and c, or, to put it another way, the 'next' cardinal number after $\aleph_0$ is c. Cantor tried to prove the continuum hypothesis, but after many years he was unable to do so. Believing it false, Kurt Gödel in the 20th century could not disprove it either, but later on he and his American colleague Paul Cohen showed that the continuum hypothesis could be proved neither true nor false within the standard framework of set theory. This amounted to saying that the continuum hypothesis was independent of the usual set of axioms that describe set theory.

This was a watershed. It was known that there were many different *types* of geometry, and now, with set theory, the independence of the continuum hypothesis brought to light the possibility of different set theories.

A host of infinities

Cantor made other discoveries on the nature of the infinite, most famously the creation of the *transfinite* cardinal numbers. Starting off with $\aleph_0$ he constructed a set with a higher order of infinity. Starting with N he constructed the set N_1 of *all* subsets of N and proved this had a higher infinity than N. Its cardinal number is written $\aleph_1$ and so $\aleph_0 < \aleph_1$. Repeating this construction on N_1 produces N_2 with cardinal number $\aleph_2$, and this can be repeated on and on, producing a whole sequence of 'alephs' of ever-higher orders:

$$\aleph_0 < \aleph_1 < \aleph_2 < \ldots < \aleph_n < \ldots$$

By showing how the sequence of alephs could be constructed, each one representing an order of infinity, Cantor had shown us how to construct an infinite sequence of infinities. A scheme of arithmetic was now possible in which cardinal numbers could be added and multiplied, yielding a rich playground for logicians and mathematicians.

We started off wondering about the nature of infinity and how 'big' it was. 'Infinity' embraces unimaginable vastness, but it also embraces an idea of imperceptible smallness, encapsulated when we describe something as infinitesimally tiny. In mathematics this is reflected in our knowledge that the whole numbers stretch off endlessly, and, at the other end of the scale, that we can go on forever creating smaller and smaller fractions of the number 1.

In probing infinity further, however, we have discovered there are sets of numbers vastly more numerous than the whole numbers. More than this, we have seen that any notion of infinity may be topped by a higher one. We have arrived at the possibility of not one notion of infinity but an infinity of them.

WHERE DO PARALLEL LINES MEET?
The birth of new geometries

*T*he short answer to the question 'where do parallel lines meet?' is that they never can — because parallel lines are defined as those lines that do not intersect one another. This, one of the most fruitful mathematical definitions ever set down, stemmed from Euclid of Alexandria, writing in 300 BC. But the story does not end there. Attempts to provide a proof for the nature of parallel lines exploded into a range of different geometries. They have changed our idea of mathematics itself and given us the means for investigating the physics of space.

For over 2000 years, until the end of the 19th century, Euclid's *Elements* was synonymous with geometry. If you wanted to study geometry you studied Euclid's account of it, and if you worked your way through the *Elements* you were assured of learning geometry from the highest authority. There was only one geometry, and Euclid had captured it in his 'sacred book'. As with the Bible, its adherents could quote it chapter and verse.

More than this, the structure of the 13 books of the *Elements* was laid out in strict logic and it became a model of how mathematics should be done: you defined your terms, laid out your assumptions, and systematically built up a body of knowledge with the theorems proved. Euclid used the work of other mathematicians, but his genius lay in this logical and coherent organization, in which, for example, Pythagoras's theorem concerning triangles was simply Book 1, Proposition 47, or just 1, 47.

The mystery of the parallel postulate

Every proposition in the *Elements* was rigorously proved on the basis of five 'postulates', that is, the assumptions from which Euclid constructed, as he thought, invincible arguments. More specifically, the postulates stated the basics of points, lines and angles, and how they work together.

The first four postulates are:

1 A straight line can be drawn from any point to any point.
2 A finite straight line can be extended continuously in a straight line.
3 A circle can be constructed with any centre and radius.
4 All right angles are equal to one another.

So far, so straightforward. There was a nagging doubt, however, about Euclid's fifth postulate, rather more elaborately worded as the 'parallel postulate':

5 If a straight line falling on two straight lines makes the interior angles on the same side less than two right angles, the two straight lines, if produced indefinitely, meet on that side on which are the angles less than the two right angles.

Euclid needed a notion of how parallel lines could be handled, and this is what forms the fifth postulate. But could it be proved from the other four postulates? If proved, it could be removed from what we have to assume. Many people tried and fell short; but the search would ultimately change geometry and transform our understanding of the physical world.

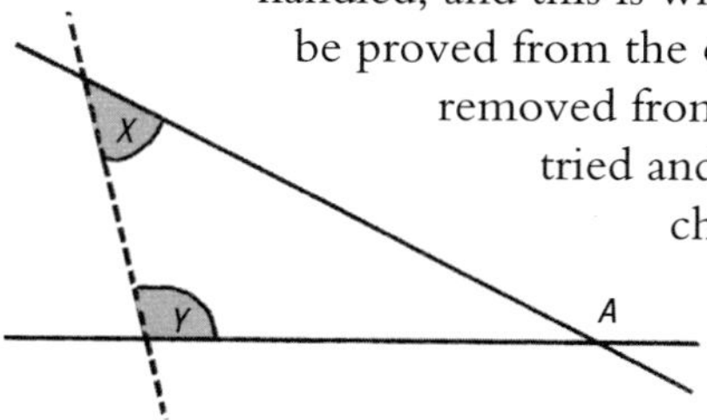

EUCLID'S PARALLEL POSTULATE

Whereas the first four postulates are short and direct, the fifth one sounds more like an essay, in which Euclid does not so much say what parallel lines are, as say what they are not. It can perhaps be better illustrated diagrammatically, in which a dotted falling line intersecting two solid lines creates angles X and Y, which, together are less than 180° (two right angles), and so the solid lines must eventually meet at a point (A) on the same side as these angles.

We might wonder why Euclid included such an awkward postulate in his logical scheme; but it is a measure of his genius that he did so. It was necessary to his geometrical arguments, but he only introduced it at the point where he was boxed into a corner and was forced to use it. He proved as much as he could without it, and its first appearance in the *Elements* is not until the proof of *Proposition 29*:

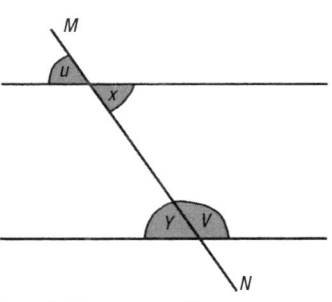

EUCLID'S PROPOSITION 29

A straight line [MN] falling on parallel straight lines makes the alternate angles equal to one another [X = Y], the exterior angle equal to the interior and opposite angle [U = X], and the interior angles on the same side equal to two right angles [X + V = 180°].

Many centuries later, the Edinburgh-based mathematician John Playfair, who published an edition of the *Elements* in 1795, gave an equivalent but more intuitive formulation of the parallel postulate:

Through a given point exactly one line can be drawn parallel to a given line.

The simplicity of Playfair's postulate gained it popularity, and compilers of later editions of the *Elements* saw it as a way of reducing the severity of Euclid's style. But whether in the language of Euclid or Playfair, the postulate still lacked a satisfying mathematical proof.

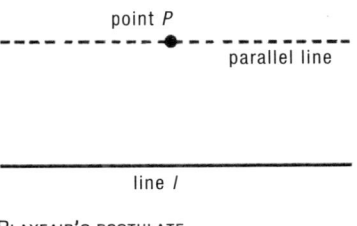

PLAYFAIR'S POSTULATE

The search for a proof

The quest for a proof of the parallel postulate tripped up many of the greatest mathematicians over the years. One of the potential pitfalls was falling prey to a kind of circular thinking, attempting to prove the postulate by ultimately appealing to the postulate itself; that is, assuming what had to be proved.

Euclid had proved that the sum of the angles in a triangle is 180°, though the proof given by Pythagoras is more transparent. If we are presented with a triangle, we can construct by the direct application of Playfair's postulate the *unique* parallel line through the 'vertex' (point opposite the base) at *A*. Using the result of Euclid's Proposition 29 about alternate angles, we

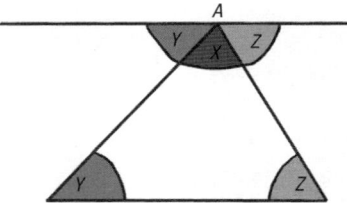

ANGLES OF A TRIANGLE = ANGLES ABOUT A POINT

mark the angles *Y* and *Z* in the diagram, and the last remaining angle *X*. We can now look at these angles in two ways. One perspective reveals that they make up the angles about a point on a line so $X + Y + Z = 180°$; the other perspective shows that $X + Y + Z$ is the sum of the angles in the triangle. QED.

The *converse* statement is also provable: if the sum of the angles of a triangle is 180°, and a line is drawn through the vertex *A* making angles *Y* and *Z* with the sides of the triangle, that line is parallel to the base of the triangle.

When a result and its converse are both proved, they are *equivalent* statements. The importance here is that Euclid's parallel postulate is equivalent to the statement that says the sum of a triangle's angles is 180°. Anyone attempting to prove the fifth postulate would have to be aware of its equivalent forms, because if any of them were used in a proof, the argument would become circular and effectively fail. The subtlety of the postulate can be seen in the cloaked nature of its equivalent statements; they appear camouflaged, not even mentioning the word 'parallel':

In any right-angled triangle, Pythagoras's theorem is true.

There is a four-sided figure in which each of its angles is a right angle.

Proving the parallel postulate became an obsession for Adrien-Marie Legendre, for more than thirty years. A thinker of the first rank, his textbook *Eléments de géométrie* (1794) was still being used a century later. But he fell into the trap of circularity by assuming the angle sum of a triangle to be 180°.

Euclid 'vindicated from all defects'

In the effort to solve the question, the efforts of the 18th-century mathematician Giovanni Saccheri are significant. He argued that in thinking of lines passing through a given point in relation to a given line there are logically just three cases to consider:

1 Through the given point there is *exactly one line* that is parallel to a given line (this is Euclid's geometry).
2 Through the given point there is *more* than one line that is parallel to a given line (that is, parallel lines are *not* unique).
3 Through a given point there are *no* lines that are parallel to a given line (that is, parallel lines do not exist).

With the second assumption, replacing Euclid's parallel postulate, Saccheri arrived at some strange theorems, which he could not explain and had never seen before. With the third assumption, he produced some contradictory results. To him, the only way out of these aberrations was to accept the first assumption, the parallel postulate. He confidently published this conclusion in his book *Euclid Vindicated from All Defects* (1733); but he had not provided a cast-iron proof.

From one parallel line to many

In the 1830s a young Russian mathematician, Nicholas Lobachevsky, came to a conclusion: the greatest mathematicians had been unsuccessful in proving the parallel postulate, so why should there not be *other* geometries, where the parallel postulate was replaced by another?

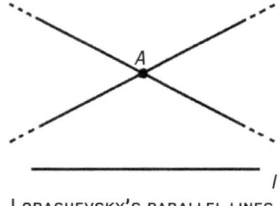

LOBACHEVSKY'S PARALLEL LINES

Lobachevsky chose Saccheri's second postulate, which proposed that parallel lines are *not* unique. This means that if we are given a line (l) and a point (A) not on this line, there is more than one line passing through the point that does *not* meet the given line. This seems, in a commonsense way, impossible. Any diagram drawn on a flat piece of paper showing more than one line through a given point surely cannot produce more than one line parallel to an existing, given line. At some stage, any other lines must surely intersect the given line. Nevertheless, around the same time the Hungarian János Bolyai also published results at variance with Euclid's treatment, and he agreed with Lobachevsky's conclusion.

The key to this thinking lay in looking beyond a flat plane to a curved surface. It was not a sphere which had a constant positive roundness, or curvature, but a surface with a constant negative curvature. Lobachevsky and Bolyai had, in effect, created the geometry of the 'pseudosphere', or, to give it its more accepted name, hyperbolic geometry. The lines on a pseudosphere are 'geodesics', that is, the curved lines of shortest length that join two points. This shape, due to Eugenio Beltrami, provided a real model for a geometry that was not Euclid's.

A PSEUDOSPHERE AND ITS MANY PARALLEL LINES (LOBACHEVSKY'S GEOMETRY)

The theorems of the new geometry did not contradict all of Euclid's theorems, for Lobachevsky and Bolyai kept Euclid's other four postulates. Theorems proved without recourse to the parallel postulate were still maintained (those up to Proposition 29). But one matter that was affected was the sum of the angles of a triangle. In Euclid's geometry, as we have seen, the fact that the angle sum in a triangle is 180° is an equivalent of his parallel postulate. When the new postulate of Lobachevsky and Bolyai was substituted, the angle sum of triangle could not equal 180°.

In fact, rather startlingly, it was rigorously proved that in this new geometry the angle sum in a triangle inscribed on a pseudosphere was *less* than 180°. In addition, the angle sum of a triangle depended on the *area* of that triangle; moreover, triangles with a larger area acquired a smaller angle sum, and vice versa. This relationship was utterly alien to Euclid, for whom the area of a triangle and the angle sum were unrelated. Any size of triangle in Euclid's geometry has the same angle sum of 180°.

Lobachevsky, a professor at the University of Kazan, and Bolyai, an army officer, published their results in the early 1830s. Instead of recognition for the magnitude of their accomplishment, their works were ignored. Here, the great Carl Friedrich Gauss came to the rescue – albeit posthumously. Unknown for his work in this field, Gauss had come to similar conclusions but feared the criticism of those who revered Euclid. So he did not publish his results. But when he died in 1855 and his correspondence on non-Euclidean geometry was published, his prestige ensured that mathematicians would also read the connected papers of Lobachevsky and Bolyai.

From one parallel line to none

Euclid's omnipotence had received a blow, but worse was to come. Later in the 19th century Bernhard Riemann made even more radical proposals than Lobachevsky and Bolyai. He changed Euclid's second postulate, 'a finite straight line can be extended continuously in a straight line', to one that said 'all lines have finite length but have no end'. Despite the apparent mystery of this statement, a circle, which in some geometries is regarded as a 'line', fits the bill. Its length, which is its circumference, is surely finite, and of course a circle has no end in the way that an ordinary line segment does.

Riemann went further. He also modified the parallel postulate, in effect choosing Saccheri's third option, that 'through a given point there are *no* lines which are parallel to the given line'. How did Riemann make parallel lines disappear? While

Lobachevksy and Bolyai alighted on the pseudosphere as the model for their new geometry, Riemann focused on a real sphere.

Spherical geometry was not in itself new – it had been studied by the Ancient Greeks and had long been important for navigation, since our world is a sphere. In Riemann's geometry the 'great circles' on the surface of spheres, those circles whose centres are located at the centre of the sphere, are the lines. The lines of longitude are lines in this geometry, but the lines of latitude are not.

To see how the parallel postulate is interpreted in this geometry, we can thus conceive of the equator (a 'great circle') as our given line and imagine a point somewhere else on the globe, say Paris, as the point P not on this line. All the lines ('great circles') through P intersect the equator at two points, and in this geometry we cannot find any line on the globe that runs through P that does *not* intersect the given line. Parallel lines do not exist in this geometry.

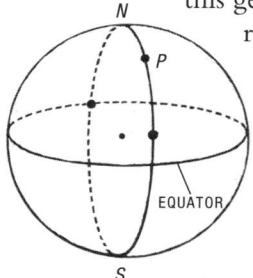

THE GLOBE: 'NO PARALLEL LINES'

The theorems proved by Euclid that did not use his Postulate 2 or Postulate 5 still held true for Riemann's spherical geometry. But it yielded more surprises for Euclidean orthodoxy. If we look at any triangle formed on the globe by the North Pole N with the equator as its base line, we find that the two base angles of triangle NAB are right angles. As there is an angle at the North Pole, in Riemann's geometry, the angle sum of a triangle therefore *exceeds* 180°.

As in Lobachevsky's and Bolyai's geometry, the angle sum depends on the area of a triangle; but here, a triangle with a larger area has a greater angle sum. If in the diagram the angle at the North Pole were increased and thus the angle sum of the triangle increased, the area of the triangle would obviously increase.

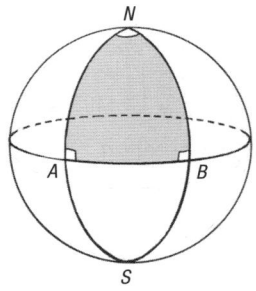

THE GLOBE: THE ANGLE SUM OF A TRIANGLE EXCEEDS 180°

A broader historical view in fact shows that the non-Euclidean geometry that Riemann was advancing had been staring man in the face for centuries. Menelaus of Alexandria had actually carried out calculations in this 'elliptical' geometry around AD 100. But man is sometimes a flat-earther and, despite the knowledge that the world is round, *locally* the world appears flat and is usually experienced as 'flat'.

> *'The cowboys have a way of trussing up a steer or a pugnacious bronco which fixes the brute so that it can neither move nor think. This is the hog-tie, and it is what Euclid did to geometry.'*
>
> E.T. BELL,
> *The Search for Truth* (1934)

Geometry *à la carte*

The fact that the new geometries were free of contradiction meant that the parallel postulate was independent of the other postulates. No one would ever be able to prove it, and any supposed proof of it would have to be wrong. Mathematicians had demonstrated their freedom to choose substitutes for the parallel postulate and derived different kinds of geometry as a result.

The Euclidean and new non-Euclidean geometries could be studied separately, but the German mathematician Felix Klein saw a useful way of ordering them that demonstrated their relationships.

	Kind of geometry	Parallel lines	Triangle angle sum
Lobachevsky, Bolyai	Hyperbolic geometry	Any number of parallel lines exist	Less than 180 degrees
Euclid	Euclidean geometry	One parallel line exists	Equal to 180 degrees
Riemann	Elliptical geometry	No parallel lines exist	Greater than 180 degrees

KLEIN'S CATEGORIES OF GEOMETRY

What mathematicians needed now, though, was a way to study spaces that, locally, conform to Euclid's geometry but which, in the bigger picture, are more complex. In 1854 Riemann gave a lecture entitled 'On the Hypotheses on which Geometry is Based', in which he introduced the concept of the 'manifold', a complex space that nevertheless, under the metaphorical microscope, appears flat (see *What Shape Is the Universe?*).

The geometry of the universe

The advent of this new 'Riemannian geometry' would be important for the analysis of the way time and space are bound together. The manner in which the presence of huge masses could alter the geometry around them required the concept of the 'manifold', whereby the measurement of length, angle and curvature could vary from point to point, unlike a sphere where the curvature is the same all over. Albert Einstein, unsurprisingly, owed much to Riemann's conceptions, saying that if he had not been acquainted with them 'I never would have been able to develop the theory of relativity'.

For over two millennia before Einstein, Euclid's *Elements* had been an extraordinary, and then extraordinarily durable, achievement. His parallel postulate was, eventually, challenged sufficiently strongly to give birth to new geometries. But they did not render Euclid's geometry 'wrong', although from a modern perspective the logic makes undeclared assumptions. From a practical point of view, we can still go on with architectural enterprises and building bridges using Euclid. Our everyday understanding of parallel lines remains 'Euclidean', and when we deal with flat surfaces we confidently head to the great geometer. But the tale of the parallel postulate has led to new geometries. Riemannian geometry and the concept of the 'manifold' are vital in man's attempts to properly understand geometry and to discover the true nature of physical space.

WHAT IS THE MATHEMATICS OF THE UNIVERSE?

The Calculus miracle

*T*he 18th-century writer Alexander Pope wrote 'Let Newton be! And all was light' in homage to a great compatriot. He might also have included the German Gottfried Wilhelm Leibniz in his praise, for the pair of natural philosophers and mathematicians pioneered the method of calculation known simply as the Calculus, which opened up accurate means of astronomical calculations and much else. Refined and honed over the years, it has proved invaluable across all the sciences. Its applications are ever widening, to encompass disciplines as diverse as demography, economics, medicine, statistics and quantum physics.

The Calculus was discovered as a result of man's looking upwards and his desire to understand the mechanics of the solar system – and beyond. In the hands of Newton and Leibniz it was successfully employed in calculating and describing the motion of the planets. Since their time it has been placed on a rigorous footing and has found applications wherever mankind has desired to describe, measure and understand *change*.

Newton the mathematician is also Newton the scientist, experimenter and astronomer. As a young man he was fascinated by planetary motion and the scientific instruments that could be used to view the heavens. When he was just 30 years old his design of a reflecting telescope caused a sensation in London and it was on this basis that he was elected to the Royal Society.

From the nature of 'centripetal' force, the force experienced by a body attracted to a point, Newton set down his great theory of universal gravitation and his famous three laws of motion. From these he used the Calculus to deduce that planets moved in elliptical orbits around the Sun. Johannes Kepler had postulated his theory of elliptical planetary motion from experimental data. Newton went one stage further and deduced it from the mathematical theory of 'change'. In the Calculus, Newton had found a mathematical key to the universe.

A world in flux

Underlying the Calculus – and by convention, the term always carries the capital letter – is the perception that the pattern of life around us is fluid, unfixed, on the move. Every minute is different from the previous one; the world is in a state of flux. In the fifth century BC, the philosopher Heraclitus summed it up: 'everything flows, nothing stands still' was his watchword, adding, for good measure, 'you could not step into the same river twice'.

We know this. From the height of a ball thrown in the air, to the distance a rocket travels from the Earth, to the size of a population, to the growth of an epidemic, to an electrical current in the presence of a magnetic object – just about any phenomenon is subject to change in space and time. The essential idea of the Calculus is that any quantity by which we measure change depends on the value of *variables*. To put it another way, the height of a ball thrown in the air depends to a large extent on the time that elapses following its release. (It is true that its height may depend also on other factors, such as air resistance, but in practical considerations some factors may be sidelined.)

The obvious variable in considering change is time; but the Calculus can handle all sorts of variables in all sorts of situations. Problems that might be considered as Calculus problems can be traced back to the Ancient Greeks, but Aristotle and Heraclitus examined change in philosophical terms rather than mathematically. Galileo and some medieval mathematicians did start down the mathematical track, but the subject really began in the modern era with those two luminaries Newton and Leibniz. They are given the credit for being the 'first' to discover the Calculus, but in truth both came upon its facets independently and with different forms of mathematical notation. Newton's notation is useful in problems where motion is involved, but it has proved less popular than the notation of Leibniz, which is now the most favoured.

The Calculus that the two men devised embraces two complementary aspects: 'Differential Calculus' and 'Integral Calculus'. Effectively, they form two sides of the same coin, an inverse relationship. The former is concerned with 'taking apart', while the latter is about 'bringing together'.

Measuring change

The central purpose of Differential Calculus is to measure the *rate of change* – how fast or slow change occurs, and this is known as the 'derivative'. The process of finding the derivative is called 'differentiating'.

Moreover, Differential Calculus is interested in the rates of change at particular moments – that is, *instantaneous* rates of change. It is thus very different from average change. If, for example, the distance from Town A to Town B is 330 miles and a car takes 11 hours to drive it, the average journey speed is 330 divided by 11, producing a figure of 30 miles per hour. But this does not tell us the speed at any particular moment during travel. At one instant the car might be stationary, at traffic lights; at another, it might be roaring along the motorway at 70 miles per hour.

Algebraically, the quantity we are measuring might be denoted as y – in the case of a ball thrown in the air, this would be its height above the ground. The rate at which y is changing is *the* key question in Differential Calculus.

Newton wrote about *instantaneous velocity*, the rate of change of distance at a particular instant; he also considered the extended or high-order concept of *instantaneous acceleration*, which is the rate of change of velocity. To focus on instantaneous change Newton thought in terms of flow. In his language, if y flowed in time he called it a 'fluent'. The *rate* at which the fluent changed he termed the 'fluxion of y' (in his notation, $\dot{y}$). Going one stage further, the rate of change of $\dot{y}$ was written as $\ddot{y}$. Thus, if y is distance, then $\dot{y}$ is velocity (or speed), and $\ddot{y}$ is acceleration.

However, Leibniz's notation for the Calculus helps us understand instantaneous velocity better. For the ball thrown in the air, we must imagine two points in time: a particular moment of time (x), which is the time at which we wish to know the velocity; and then another moment of extra time, later on. The time between these two moments is denoted as Δx. (It is *one* symbol, not $\Delta \times x$, and it is vocalized as 'delta x'.) During this extra period of time, the ball will have travelled a distance written as Δy ('delta y').

The *average* speed over this extra time — the distance travelled divided by the time taken — is expressed as $\frac{\Delta y}{\Delta x}$. The smaller the interval of time represented by Δx, the closer $\frac{\Delta y}{\Delta x}$ will be to the *instantaneous* speed at the moment x.

Factoring in acceleration

To appreciate better how the nuts and bolts of the Differential Calculus fit together, consider the example of a rocket being launched. It is similar to a ball being thrown in the air, but we have to contend with the nature of rocket propulsion too.

So, we want to figure out the instantaneous speed of the rocket after x minutes. If the speed were constant — say, 10 miles per minute — it would be *uniform* motion and this would also be the instantaneous speed, at any moment. But the motion of a rocket is not like that. It begins from nothing, moves slowly at first and gathers more and more pace as it accelerates.

In reality, the relationship between distance travelled and time taken would depend on all kinds of rocket-related factors, such as the amount of fuel on board, the air resistance and the reduction in the pull of the Earth's gravity as the rocket ascends into space. However, if we conceive of the miles travelled as y and the minutes taken as x, then let us assume a relationship in which $y = x^2$. This equation may be over-simplified, but it possesses the essential characteristics of rocket motion in that it embraces acceleration: it would mean that the rocket has travelled 1 mile after 1 minute, but 25 miles after 5 minutes.

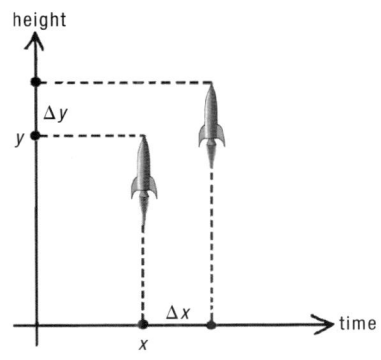

ROCKET VELOCITY MEASURED USING THE CALCULUS

In order to compute the instantaneous speed when the rocket is 5 minutes into its flight, we need first to calculate, using Leibniz's notation $\frac{\Delta y}{\Delta x}$, the average velocity during our short extra period of time Δx.

Let us assume that $\Delta x = 0.01$ minutes (that's less than one second), so that means our two points of time (x) are 5 minutes and 5.01 minutes. Using our assumption that $y = x^2$, we can work out that the distance (y) travelled during this period is 0.1001 miles, a calculation arrived at by subtracting 5^2 from 5.01^2.

This means the average velocity $(\frac{\Delta y}{\Delta x})$ over this interval of time can now be calculated as 0.1001 (miles) divided by 0.01 (minutes), producing a result of 10.01 miles per minute. Over even smaller time intervals – that is, by applying a limiting process, the basic idea behind the Calculus – the *average velocity* gets closer and closer to an *instantaneous velocity* of 10 miles per minute at the point when the rocket has been in flight for 5 minutes (x).

In fact, it emerges that, on the basis of our assumption, the 'expression' $y = x^2$, *whatever* value we choose for time x, the instantaneous velocity will still equate to a number that is a doubling of x. Leibniz presented this conclusion in his notation as $\frac{dy}{dx} = 2x$, in which $\frac{dy}{dx}$ is the *derivative* of $y = x^2$. Leibniz thought of his notation as being an infinitesimal dy divided by an infinitesimal dx though current mathematical thinking insists that the derivative $\frac{dy}{dx}$ is one symbol.

Finding the derivative – 'differentiating' – lies at the heart of the Differential Calculus. It can be calculated for other mathematical expressions too, such as $y = x$ in which case $\frac{dy}{dx} = 1$, and $y = x^3$ in which case $\frac{dy}{dx} = 3x^2$.

The other side of the coin

With our rocket, we assumed an expression for the distance travelled by the rocket and, using the Differential Calculus, calculated its velocity. But the other side of the coin, Integral Calculus, provides the means for solving the inverse type of problem: we may know the speed of the rocket, but we now want to figure out the distance it has travelled at a particular moment. This depends on locating the *integral* of velocity, the process mathematicians call 'integrating'.

If we first suppose that the rocket has a *constant* velocity (v) of 10 miles per minute, and we draw we draw a graph of velocity against time, it will just have a straight line passing through $v = 10$, describing the fact that at any time (along the x time axis) it has this speed.

What is its height after 5 minutes? As it travels at constant speed of 10 miles per minute the height is simply $5 \times 10 = 50$ miles. From the representation as a graph, we can see that the distance can be interpreted as the *area*, which is also equal to 5×10. In other words, for the inverse problem of Integral Calculus, we calculate *area* in order to find height. Because of acceleration, of course, our rocket is not moving upwards at a constant speed. So let's assume the velocity v is changing, and at any moment its miles per hour are equivalent to twice the amount of time travelled, that is, $v = 2x$. This proposes that the rocket travels faster as time x advances. Now, what happens if we draw a corresponding graph of $v = 2x$?

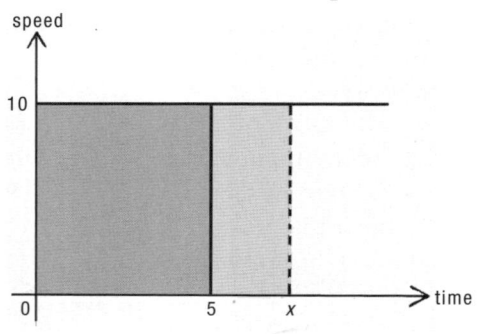

DISTANCE TRAVELLED, REPRESENTED AS AREA, WITH UNIFORM SPEED

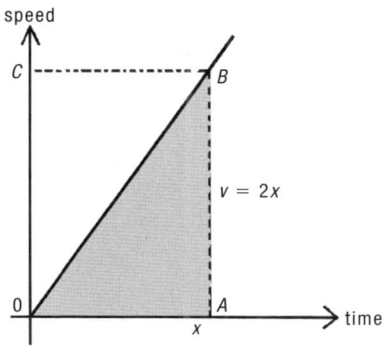

As before, the distance travelled will be given by the area, here the area of the triangle represented by the corners OAB. This, as the graph shows, is half of the area of the rectangle whose corners are $OABC$. So the calculation to derive the distance travelled involves $OA \times OC$, which is $x \times 2x = 2x^2$. The area of the triangle is just half of this, that is, x^2. This is the distance the rocket has travelled if the velocity is based on the assumption $v = 2x$.

DISTANCE TRAVELLED, REPRESENTED AS AREA, WITH ACCELERATION

Finding the area under a graph is the central idea of the Integral Calculus and also a motivation for Newton and Leibniz. Taken by itself, the process of finding the area under a graph is one of summing up small rectangles, of area vdx, and historically the word 'sum' was used. When this was abbreviated to the single letter S, it became formalized in Leibniz's mathematical notation as the elongated $\int vdx$ and this is called the integral of v. We have found that $\int 2xdx = x^2$ or equivalently (dividing by 2) that $\int xdx = \frac{x^2}{2}$.

As for the derivative, integrals can be calculated for other mathematical expressions too, for example $\int x^2 dx = \frac{x^3}{3}$ and $\int x^3 dx = \frac{x^4}{4}$.

Differential–Integral relations and the 'limit'

So how do Differential and Integral Calculus come together productively? Given the height (y) of our exemplary rocket we were able to work out its velocity – Leibniz's $\frac{dy}{dx}$ – using the Differential Calculus, and, conversely, given the velocity (v), we found out its height by calculating $\int vdx$ using the Integral Calculus. In mathematical terms, the related nature of our examples can be shown in their equations:

Differential Calculus (to find velocity): Given $y = x^2$ we found $\frac{dy}{dx} = 2x$

Integral Calculus (to find distance travelled): Given $v = 2x$ we found $\int v dx = x^2$

The two operations of differentiating and integrating are inverse operations, as both Newton and Leibniz recognized. (Indeed, the integral is sometimes known as the *anti-derivative* to emphasize the relationship.)

The Calculus did not, of course, emerge fully formed at its Newtonian–Leibnizian birth. It came into the world strong but ungainly. What was needed to shore up the foundations was another concept – the *limit*. In the event, this arrived only after Newton and Leibniz, and after many leading mathematicians had failed to make Calculus rigorous. The nettle was grasped by Augustin-Louis Cauchy in the 19th century, who perfected a notion of limit and set the whole theory of the Calculus on a more secure basis.

The Calculus and optimization

A leading application for which Calculus has proved invaluable is finding a desired minimum or maximum value of some quantity. We can appreciate this by way of another example, this time earthbound rather than in the skies.

Let's assume that a farmer has 800 metres of fencing and wishes to enclose a rectangular area of land that abuts a straight river. He wants to discover how he can use his finite quantity of fencing to enclose the largest area of land, for maximum benefit and profit. It is obvious that using the river as one of the long sides of the rectangle, rather than as a width, will play to his benefit. Then he might try to solve the puzzle by drawing pictures or by experimenting with a variety of widths and lengths that can be accommodated by his 800 metres. But Calculus would enable him to do away with guesswork and move straight to his solution.

If the farmer defined the width of his rectangular field as x and the single length of this l, then the length l of the rectangle in metres could be expressed as $800 - 2x$.

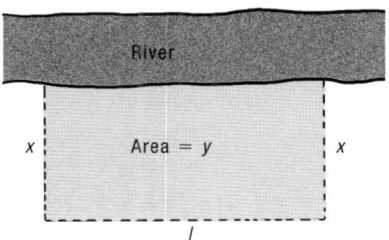

The area of the field, equivalent to $x \times l$, can now be rephrased as $y = x \times (800 - 2x)$ which in turn equates to $y = 800x - 2x^2$. If, for example, the farmer chose the width $x = 100$ metres, the area y would then be $100 \times (800 - 200) = 60,000$ square metres.

THE FARMER'S FIELD BOUNDED BY A RIVER

But could he do better and locate a value of x that delivers a larger field? We'll now take the problem out of the farmer's world and into the mathematician's world. We can show the varying size of the area y with different values of x by representing the above equation in a graph. In this, as width x increases from zero, the area y increases to its maximum value until it reaches a point where it begins to diminish again: the graph is a curve.

We have encountered the derivative $\frac{dy}{dx}$ in terms of the speed of the rocket, but another valuable interpretation of it is as the gradient to a curve. The steepness at a point on a curve means the gradient of the tangent line at that point. This gradient is approximated by $\frac{\Delta y}{\Delta x}$ and is exactly equal to $\frac{dy}{dx}$.

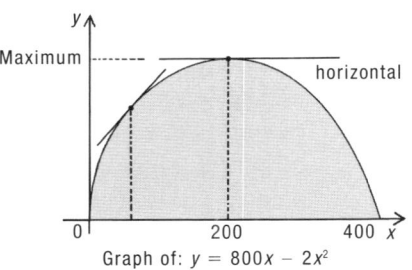

Graph of: $y = 800x - 2x^2$

MAXIMIZING AN AREA USING THE CALCULUS

At the highest point on the curve, the point we want, the steepness is zero because the tangent is horizontal. So if we can calculate the derivative of y, by setting it equal to zero, we automatically have an equation in the unknown x.

The derivative of $800x - 2x^2$ is $800 - 4x$ and by solving the equation $800 - 4x = 0$ we fnd $x = 200$ is the optimum value, yielding an area of 80,000 square metres.

From equations of nature to mathematical modelling

So-called 'equations of nature' are equations where some of the terms are themselves derivatives, that is, rates of change, so they are conceived of as *differential equations*. An example would be the proposition that the derivative $\frac{dy}{dx} = 3x^2$, where the problem is to find y so that the left- and right-hand sides of the equation are equal.

The field of differential equations is huge, and besides mathematicians it attracts physicists involved with physical theories, chemists interested in the rate of chemical reactions, biologists interested in diseases and how fast they are spread. These are studied within the framework of mathematical modelling, where simplifying assumptions are made in order to understand a process. Many areas where the Calculus is applied involve quantities with more than one variable, such as space *and* time. For these, mathematicians have taken recourse to the notion of a *partial derivative*, which is obtained by taking each variable separately. That brings us into the realm of *partial differential equations*, of which there is a whole range of classical examples with real-life applications. They include the Navier–Stokes equations, elaborated by two men in the 19th century, which are used for weather forecasting and the motion of fluids such as ocean currents; (James Clerk) Maxwell's equations, also from the 19th century, for linking electric and magnetic fields; and Albert Einstein's field equations in general relativity.

Partial differential equations have found applications in economics too, one example being the Black-Scholes equation that is used to try to predict stock prices, originated by the US economists Fischer Black and Myron Scholes. Related to the classical equation regarding the distribution of heat, their work gained its authors the Nobel Prize for Economics in 1997.

Calculus is the enduring legacy of both Newton and Leibniz. There is hardly a corner in science, social science, statistics or any field of engineering that does not owe a debt to its magnificent 17th-century originators and their musings on the elliptical motion of the planets.

ARE STATISTICS LIES?
Data, proof and 'damned lies'

*T*he *pioneer pollster George Gallup famously remarked that he could prove God exists by means of statistics. The sceptical 19th-century historian Thomas Carlyle equally famously excoriated statistics for their ability to prove anything that one might desire. These were extreme but attention-grabbing claims. But like it or not, today we live in a world saturated in statistics, as armies of researchers go about the business of collecting data and drawing inferences from them. But what mathematical means do they employ, and what, if anything, can be established with statistics?*

Statistics have often suffered from a bad press, being regarded with suspicion as a specious means of arguing a point, especially in the realm of politics – the field from which the term 'statistics' emerged. Mark Twain attributed the accusation 'lies, damned lies, and statistics' to Benjamin Disraeli, and while it is doubtful if the Victorian statesman actually used the phrase, it has since been uttered many times to express contempt for dubious arguments 'reinforced' by data.

But is this reputation justified? Certainly, and despite it, modern society and economy could not function without statistics: they underpin the decisions of governments, national and international organizations, and medium and large companies.

Statisticians are, correspondingly, highly sought after professionals. And there is something comforting and non-abstract about mathematical practices that appear so very much tied to practical applications.

Pie charts and the Lady of the Lamp

The relationship between statistical figures and the goal of advancing an argument means that presentation plays an important role. Statistics generate a panoply of diagrams and charts to translate often intimidating amounts of numerical data into digestible and persuasive visuals. This lesson was learned early on, including, in the 1850s, by Florence Nightingale, who administered hospital care for the British Army in the Crimean War. She marshalled statistics to highlight the poor conditions in which British soldiers had to serve, claiming that more lives were lost through disease than by combat and wounds. She designed a form of pie chart, which she called her 'bat's wing', to dramatically illustrate the causes of mortality.

In her diagrams, which chart deaths per month, the causes of mortality are arranged in three colour-coded categories: preventable disease, combat-related wounds, and other causes, and the areas of the sectors are proportional to the number of deaths. Plainly, the preventable disease category was the largest sector in every month. The diagrams were intended to *persuade*, 'to affect through the eyes what we fail to convey to the public through their word-proof ears'. In this case, the statistics were sober reading indeed, and their visual representation potently encapsulated them. Nightingale's contemporary, Dr John Snow, also brought a form of statistical presentation bear when he plotted cases of cholera on a map of London. The patterns of distribution that became apparent reinforced his conclusion that the disease

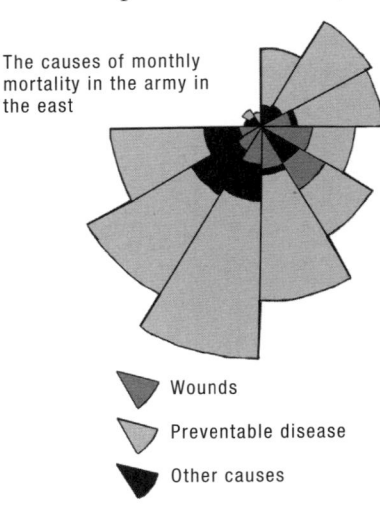

The causes of monthly mortality in the army in the east

Wounds

Preventable disease

Other causes

FLORENCE NIGHTINGALE'S 'BAT'S WING' CRIMEAN WAR CHART

was spread by foul drinking water from the public pumps rather than from, as was widely believed, 'bad air'. While both approaches used statistics in a very simple way, they both demonstrated their points.

The laws of averages

In the documentation of the Crimean War and London's cholera epidemics, the totality of data was looked at in terms of distribution and causes and then represented in the resulting charts and maps. However, it is often the case in statistics that pages of data are reduced to a single important measure, such as the 'average'. But what does 'average' mean? If the average salary in a company were £45,000, it might mean that everyone earned this amount, or it might mean that most earned less but, say, the chairperson's salary of £200,000 had been included in the calculation. The latter might be more likely, but one wouldn't really know.

'In the original sense of the word, "Statistics" was the science of statecraft: to the political arithmetician of the eighteenth century, its function was to be the eyes and ears of the central government.'

R.A. FISHER, STATISTICIAN

What most people think of as an 'average' is usually the *mean*, the figure obtained by adding the readings together and dividing by the total number of them. The problem here is that an isolated high value or low value can produce an 'average' that, while technically correct, actually gives a false impression. If the average is the *median*, that is the 'middle value' when the data is arranged in numerical order, then the average would be insensitive to a very high/low value. However, if the average is the *mode*, which is the most commonly occurring value in the data, we would know the salary that most people in the company received.

In producing an average, statisticians are often also interested in the extent to which figures *deviate* from the average. At its simplest (using our example), the variation from the mean can be worked out by adding together all the individual

> *'I do not know whether there is anything peculiarly exciting in the air of this particular part of Hertfordshire, but the number of engagements that go on seems to me considerably above the proper average that statistics have laid down for our guidance.'*
>
> LADY BRACKNELL, IN OSCAR WILDE'S
> *The Importance of Being Earnest* (1895)

deviations from £45,000 in salaries (a salary of £40,000 would produce a deviation of £5,000). The resulting total of all deviations is then divided by the number of readings, to produce an average deviation. Statisticians are particularly interested in a more sophisticated calculation, the 'standard deviation', signified by the Greek letter σ (sigma). It is the statistics of deviation that can often reveal a different kind of truth behind headline figures, for example that, while a group of people might be getting wealthier as a whole, the inequalities among them may also be growing.

Sampling

Very often it is simply impractical and unwieldy to attempt to gather *all* the data for a particular phenomenon, such as a whole population: a government could hardly mount a census every time it wished to know something about its people. A vital aspect of much statistical investigation therefore involves identifying a reliable sample group and then drawing inferences from the results.

Let's imagine, by way of example, that we wish to find the average height of a population. Statisticians would take care to choose a *random* sample of people to avoid any inbuilt bias. It wouldn't make sense to sample in neighbourhoods where, say, there were above-average numbers of infants. Statisticians learned a sobering lesson from the 1936 US presidential election, when a forecast predicted a tight race, but Franklin D. Roosevelt won by a huge landslide. In this case, the pollsters had sampled from telephone directories and lists of car-owners, and had inadvertently arrived at a sample skewed towards one, relatively affluent, socio-economic group.

To estimate average height, statisticians have to figure out the population *mean*, which has a traditional symbol in μ, the Greek letter vocalized as 'mu'. If the sample group is properly chosen, then it is likely that the average height of people in the sample, which statisticians denote as $\bar{x}$ (vocalized as 'x bar'), will be a good *estimate* of the population mean μ.

Estimating in a rigorous fashion is where a vital tool in the statistician's armoury comes in: the central limit theorem, which states that, if one repeatedly takes samples, the sample means $\bar{x}$ should follow what is known as the 'normal distribution'.

The bell curve

The 'normal distribution', represented graphically, is the famous bell-shaped curve. It has been described as being as fundamental to statistics as the straight line is to mathematics. Abraham De Moivre, whose family emigrated to London from France in the 16th century, discovered its main properties, but it only acquired the designation 'normal distribution' after it had passed through the hands of others. (It is sometimes known as the Gaussian distribution, after Carl Friedrich Gauss's work in the 19th century.)

THE NORMAL DISTRIBUTION OF COINS TOSSED

De Moivre tackled a problem akin to the traditional coin-tossing problem. If we toss a coin a certain number of times, say n times, we can work out the probability of getting x heads. De Moivre considered a *large* number of tosses and found a quick route to these probabilities because they followed a normal distribution.

If we toss a coin 100 times, the mean of the distribution of heads thrown is 50. This is hardly surprising, since it matches the probability of getting heads around half the time. De Moivre showed

> *'Some people hate the very name of statistics, but I find them full of beauty and interest ... They are the only tools by which an opening can be cut through the formidable thicket of difficulties that bars the path of those who pursue the Science of man.'*
>
> FRANCIS GALTON,
> *Natural Inheritance* (1889)

us how we can measure the probability by calculating the area under the bell curve. So, for example, the probability of getting more than 60 heads means calculating the *area* under the curve to the right of 60. (In fact, statisticians are able to refer to normal distribution tables for such purposes.) We would find the probability of getting 60 or more heads is 0.0228; if we were to *repeatedly* toss the coin 100 times, we would expect to exceed 60 heads about 2% of the time.

It is the central limit theorem that links the practice of sampling with the normal distribution. The theorem says that the means, i.e. $\bar{x}$, of random samples follow a normal distribution centred about the mean μ, and that a large sample will give a more accurate answer to our estimate of μ than a small one. According to the theorem, if the variation of heights in the population is known, the variation of the sample means $\bar{x}$ is reduced by a factor of $\frac{1}{\sqrt{n}}$, so the bigger the random sample the better the estimate of the population mean μ will be. This is to use statistics as a science.

The science of statistics

Florence Nightingale's chart conveyed a message and effectively proved a point, but it did not draw on statistical theory. The science of statistics, by contrast, is the mathematical theory used to draw inferences about populations based on samples.

Furthermore, the nature of statistics is to quantify the difference between what we *expect*, based on theory, and what we *actually* get as the result of an experiment. So, in the coin-tossing experiment, we expected 50 heads in 100 tosses, and we can measure the probability of a deviation from this number. The leading British statistician R. A. Fisher formulated a method for handling this type of analysis in his *Statistical Methods for Research Workers* (1925), a practical guide in which he described 'significance tests'.

Fisher's famous example of a significance test concerned the art of pouring tea. Should tea be poured into milk already in the cup, or should milk be added to the tea after it is poured? A lady of his acquaintance claimed she could tell the difference. Acting as sceptic, Fisher set down a hypothesis that she had no such ability. He then tested her claim with a tea-tasting experiment. If she obtained eight successes in eight trials Fisher would say the result was *statistically significant* as it was unlikely that this result could be obtained by pure chance. He would then reject the hypothesis as unlikely and acknowledge there was more to tea making than he had assumed.

How would Fisher apply significance tests to the coin problem? How can we decide whether a given coin is fair? It might have been weighted, but we are not allowed to inspect it for any tampering. To set up the experiment, Fisher would frame the hypothesis that 'the coin is fair', which he described as testing the 'null hypothesis', a term derived from physicists who described a successful experiment as 'null' when it delivered no readings. It is usually denoted by H_0. So Fisher could accept H_0 or he could reject it with a high degree of confidence.

Suppose we achieved an extreme result: 100 heads on tossing the coin 100 times. Could we conclude that the coin is weighted and unfair? Not with certainty, for we might have got 100 heads by chance. However, this outcome is *very* unlikely on the basis that H_0 is correct. The probability of getting 100 heads with a fair coin is actually 0.5 multiplied by itself 100 times, producing a miniscule number, about 8×10^{-31} (that's 30 zeros after the decimal point). We might then reasonably conclude that the likelihood of the coin being fair is accordingly miniscule.

'Statistics is a science which
ought to be honourable,
the basis of many most
important sciences ... [but]
a wise hand is requisite for
carrying it on.'

THOMAS CARLYLE,
Critical and Miscellaneous Essays (1860)

Could we draw the same conclusion if the outcome had been 60 heads? In other words what amount of divergence of the experimental result from the expected number of 50 would cause suspicion? The probability that the number of heads is 60 or more, based on the coin being fair, is around 2%, a low enough probability for Fisher to question whether the coin were fair.

A competing method of testing hypotheses was put forward by Jerzy Neyman and Egon Pearson in the 1930s, one which eventually resulted in a bitter controversy with Fisher. In Fisher's theory there is only one hypothesis, and attention is focused on rejecting it. In the Neyman–Pearson theory, there are *two* competing hypotheses. Their decision model leads to consideration of an alternative hypothesis, H_1. Thus, sticking with our coins, Neyman and Pearson could hypothesize:

H_0: the coin is fair
H_1: the coin is unfair

Or, depending on the aim of the experiment, they might hypothesize:

H_0: the coin is fair
H_1: the coin is biased towards heads

In this approach, a decision rule is given *before* any experiment takes place. We don't wait until we have found something unlikely. Moreover, two probabilities of making an error are brought into the theory:

The probability of concluding the coin is unfair (rejecting H_0) when it is in fact fair (H_0 is correct).

The probability of judging the coin fair (accepting H_0) when in fact it is unfair (H_0 is false).

The Neyman–Pearson theory gained credence, because it appealed to probability explicitly and because of this effective use of the 'balance' between the two probabilities of making an error.

Fisher's disciples continue to debate these issues with the followers of the Neyman–Pearson approach. There is also a third way, advocated by adherents of the 'Bayesian' persuasion, named after the 18th-century cleric and mathematician Thomas Bayes. Bayesians like to introduce the idea of a prior probability that a hypothesis is true and limit conclusions to such statements of the form 'The probability of the coin being fair is such-and-such a value based on the data we have obtained.' The debate rumbles on.

Statistics, proof and truth

Whatever competing methods statisticians adopt to enhance the credibility of their findings, their results penetrate our lives. Governments justify their policies according to statistics of crime, immigration, employment and much else. Experimental psychologists employ statistics as they investigate perception, memory, or attention span. Market researchers are constantly asking us about our circumstances and choices, and the answers warrant vast sums from commercial companies that wish to sell us products. Sociologists and demographers perform large-scale experiments through sampling, while 'hard sciences' such as thermodynamics and some branches of mechanics also engage in statistical analysis. In astronomy, Gauss led the way in using the technique of 'least squares' to minimize observational errors. His work enabled him to analyse observational data to successfully relocate the small planet Ceres after it had been obscured by the Sun, but the 'least squares' technique is of greater value in modern statistical theory than finding a minor planet.

What, ultimately, can statistics prove? As ever, this depends on the question asked, and there is *much* that can, and has been, credibly demonstrated by statistical method. Statistics cannot, though, prove causation. Rather, it shows an association. If, returning to our examples, it turns out that taller people live in wealthier neighbourhoods, an association between income and height might be demonstrated – but it is another leap to then argue that one is the cause of the other.

However, in some cases further statistical examination can narrow, even clinch, the argument. In the campaign against smoking, a causal link between smoking and cancer was never definitively established; it was not proved that the physiological reason for the onset of cancer was caused by smoke entering the lungs. But a *statistically significant association* was established, by comparing groups of smokers against control groups of non-smokers. At a conservative estimate, it was shown that 80% of people who died of lung cancer each year were smokers, and this served as proof enough.

While some statistical efforts can misfire badly, as occurred during the 1936 US election, statisticians can also get their answers spectacularly right. Modern exit polls, taken from samples of voters after polls have closed, have often been highly accurate. In the case of the British General Election of 2010, the supposedly rogue exit polls aroused initial derision from political pundits, until the actual results vindicated the statistics.

Dry statistics do not prove anything with unbridled certainty, and they are capable, in the wrong hands, of being abused. Yet certainty is hardly ever attained in human affairs and in practice we can only draw conclusions with a certain probability of them being true. With well-designed statistical investigations, we hope to place limits on life's uncertainty. In that sense, the statistical method is very far from vacuous, and it is not a lie. In most situations, it is all we have.

CAN MATHEMATICS GUARANTEE RICHES?

Uncertainty, chance and probability theory

*H*uman beings have always wanted to narrow the odds. *In its largest, and most fundamental, sense this is the timeless effort to gain mastery over our environment and minimize the risks — failed harvests, natural disasters, the lurking accidents and misfortunes of everyday life. But there is a lighter side, too, that reflects humankind's playfulness, competitive spirit and, if money is at stake, the prospect of riches. Modern probability theory, the mathematicians' efforts to narrow the odds, started off relatively late, with games of chance played by 17th-century gamblers. But dice and cards proved a spur to a modern theory that now lies at the heart of how we evaluate risk and cope with uncertainty.*

Gambling *can* make us rich. We see this, weekly, as the results of national and international lotteries are announced, and the ranks of millionaires increase. In the world of high finance, sophisticated betting creates fortunes; and the casual punter can get lucky on the horses or from a devil-may-care weekend in Las Vegas. But winners necessitate losers. What about that horse that started as a pre-race certainty but came in last? What about those stock market crashes? What about the gambler who leaves the casino in the early hours having 'lost his shirt'?

> 'All knowledge resolves itself into probability.'
>
> DAVID HUME,
> *A Treatise of Human Nature*

The many forms of gambling offer different challenges. In the lottery, you just need luck, since past form is hardly a guide to the future. This does not apply to the dealer in stocks and shares, who scrutinizes the track records of companies. A shrewd gambler on sporting events would need to know how the games are played, the recent form of the participants, and how the odds stack up. And a Las Vegas vacationer should remember to have the sense to quit while on a winning streak.

Knowledge and intuition can be a great help, but, ultimately, is there a way to bet *sensibly*, to optimize our chances and avoid potential calamity if a lot is at stake? The first step in tackling this question is to gain a proper understanding of how odds are calculated. That means mathematics, and, specifically, the mathematics of 'probability'.

Playing with cards and dice

The archetypal, simple gambling problem is tossing a coin. If we throw a coin in the air it must come down as either heads (H) or tails (T). True, it *might* land on its edge, but the possibility is remote, and we would simply ignore the throw and throw again. We might reasonably say that the chance of getting heads is one chance in two, or, more technically exact, the 'probability of heads is $\frac{1}{2}$'. Likewise, the probability of tails is $\frac{1}{2}$, and so this equivalence makes the traditional tossing of a coin a fair way to decide on the dispositions of teams or players at the beginning of sporting events.

A curious facet of probability calculations is that they can confound intuition and expectations. We can suggest this by sticking with our coins. How many throws of the coin are more likely to give a result of two heads – three or four throws? Intuitively, we would choose the second option because, surely, it gives us more chances. If we throw the coin three times, there are $2 \times 2 \times 2 = 8$ possible outcomes. Three outcomes, HHT, HTH, THH, would give us a result of two heads, so the probability of getting two heads is therefore $\frac{3}{8}$.

Now let's throw the coin four times. In this case there are $2 \times 2 \times 2 \times 2 = 16$ possible outcomes in all, and six outcomes HHTT, HTHT, HTTH, THHT, THTH, TTHH would deliver the result we want. The probability in this case is $\frac{6}{16} = \frac{3}{8}$, the same as before! With four throws of the coin there are *more* outcomes; but the probability is the *ratio* of successful outcomes to the total number of outcomes, and this accounts for the fact that we end up with the same answer.

The 17th-century amateur mathematician Antoine Gombaud applied himself to a dice problem, wondering whether it was better to bet on getting *one six* on four throws of a dice, or getting *a double-six* on 24 throws of *two* dice. According to the popular view at the time, the second bet, of getting a double-six, was more likely because of the far larger number of throws. Could mathematics resolve the question?

The probability of *not* getting a six on a single throw is $\frac{5}{6}$, so the probability of not getting a six on four throws becomes:

$$\frac{5}{6} \times \frac{5}{6} \times \frac{5}{6} \times \frac{5}{6} = \left(\frac{5}{6}\right)^4$$

The logical possibilities are that either we don't get a six *or* we get at least one six: it is certain that one of these results will happen. Something that is certain to happen has a probability of 1, so we can say that the probability that we don't get a six *added* to the probability of getting at least one six is equal to 1. This means that the probability of getting at least one six will be 1 minus the probability of not getting a six:

$$1 - \left(\frac{5}{6}\right)^4 = 0.5177 \text{ (or approximately 51\%)}$$

If we throw two dice together, there would be 36 possible outcomes on each throw. This is because we could combine a score of one on the first dice with any score on the second dice (making six outcomes), then a score of two on the first dice with any score on the second (making another six outcomes) and so on, making 36 outcomes altogether. There are 35 of those combinations that are *not* a double-six, making the probability of *not* getting a double-six $\frac{35}{36}$. This means that the probability of not getting a double-six on 24 throws is $\left(\frac{35}{36}\right)^{24}$.

Following the same procedure as in the previous calculation for one dice, the probability of getting at least one double-six becomes:

$$1 - \left(\frac{35}{36}\right)^{24} = 0.4914 \text{ (or approximately 49\%)}$$

In other words, the chances of getting a double-six on 24 throws are slightly lower, at 49% compared with 51%.

Lottery fever

The numbers of possible permutations with a coin or a couple of dice pale in comparison with the numbers in modern national and international lottery draws. Some people have wryly observed that we have as much chance of winning this type of lottery as being struck on the head by an asteroid; nevertheless, there is usually one lucky winner, an instant millionaire, and this fact ensures that the *possibility* of winning remains tangible. But what are the odds?

'I figure you have the same chance of winning the lottery whether you play or not.'

FRAN LEBOWITZ

A good example is the British National Lottery, where punters select six numbers from the range 1–49 and, if the same six coincide with those picked randomly by a machine, the lucky ticket-holder wins the jackpot. With probability theory we can calculate the probability of the numbers coming up. For the first number, there are 49 possibilities, and, as this number cannot be the same for the second, there are 48 possibilities for the second. So there are 49 × 48 possibilities for the first two numbers combined. Carrying this on for all six numbers, the number of possibilities becomes:

$$49 \times 48 \times 47 \times 46 \times 45 \times 44$$

This calculation takes account of the *order* in which the numbers are selected, though in the lottery we don't mind about the order. Given six numbers, there are $6 \times 5 \times 4 \times 3 \times 2 \times 1$ ways of arranging them, so pulling all this together produces:

$$\frac{49 \times 48 \times 47 \times 46 \times 45 \times 44}{6 \times 5 \times 4 \times 3 \times 2 \times 1} = 13,983,816$$

This figure is the amount of combinations of six numbers; and out of them only one can be a winner. That means the probability of winning is tiny, equal to $\frac{1}{13,983,816}$ or approximately 0.00000715%!

There is a way we could *guarantee* winning. We could buy enough tickets so that we cover all the possible number combinations. At £1 per ticket, that would cost £13,983,816. But even here there would be a risk, for we do not exclusively own any number combination, and we might have to share our winnings with another holder of the winning sequence. There is never 'a sure thing' in the betting world.

The Monty Hall puzzle

Staggeringly small though the odds are for winning a lottery, their calculations of probability do appear logical. But some probability puzzles appear to defy logic. One of the most perplexing – to mathematicians and laymen alike – was inspired by Monty Hall, who bounced across US television screens in the 1960s with the show *Let's Make a Deal*, where contestants could win prizes by choosing boxes.

It runs like this. In each of three closed boxes (A, B and C) there is a piece of paper. In two of the boxes, the piece of paper is blank; in the third, it guarantees the winner a free holiday. The contestant has to try his or her luck and guess the winning box.

So, we choose a box, but Monty – who knows the contents of each box – won't let us open it immediately. Instead he opens and shows us *another* of the boxes, careful not to open the winning box, and now allows us to make a final choice. The burning question is: should we stay with our original choice or switch to the other unopened box?

One piece of advice would be 'don't switch'. Because Monty has shown one box to contain a blank sheet, we know of the remaining two boxes one contains the holiday. It is surely a 50:50 choice, so what is the point of switching? In one respect the reasoning is sound: we cannot know for sure which box contains the holiday, so we can only base our decision on our notion of which is more likely.

The paradox is that *not* switching our choice decreases the likelihood of winning. This is the reasoning. With three boxes, the probability of choosing the box with the holiday is $\frac{1}{3}$, since we have no reason to prefer one box to another.

Let's assume the holiday is in Box C (which Monty knows, of course), but we begin by choosing Box A. Monty would now have to show us Box B, for he cannot show us the winning Box C. Our choice is now between sticking with A or moving to Box C. If we switch to Box C, we get lucky. The argument would be similar if we had originally chosen Box B. In this case Monty would have shown us Box A and we would have the choice of sticking with Box B or switching to Box C, and if we do switch we get lucky. The probability of winning so far *if we switch* can be stated as: $\frac{1}{3} + \frac{1}{3} = \frac{2}{3}$

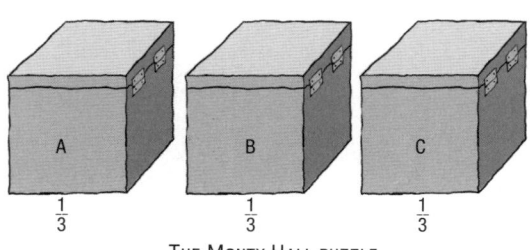

$$\begin{array}{ccc} A & B & C \\ \frac{1}{3} & \frac{1}{3} & \frac{1}{3} \end{array}$$

THE MONTY HALL PUZZLE

If, by contrast, we begin by choosing the winning Box C, Monty would have to show us Box A or B. Now, if we switch, we will lose, with a probability of $\frac{1}{3}$. Therefore, considering all the possible scenarios, we double our probability of winning the holiday from $\frac{1}{3}$ at the outset to $\frac{2}{3}$ by switching our choice. What has happened here is that Monty has given us information; we are not starting off with complete ignorance when the number of boxes has been reduced to two.

Poker players and card counters

Today's casual gambler does not need to be a TV show contestant or even head to a casino or a betting shop. The 21st century has ushered in the virtual world of online poker. Traditionally the card game of serious gamblers, poker has many variants with exotic names, but the basic poker, of the kind once played out in the Wild West, is five card draw. Here, each player is dealt five cards and given the option of discarding and receiving further cards.

Players appraise their hands in a hierarchy determined by quality. At the top is the royal flush (ace, king, queen, jack, ten, of the same suit), followed by a straight flush (five cards of the same suit in sequence), and four of a kind. A player wins when they beat the other players with the quality of their hand. At the lowest end of the hierarchy, a player has to rely on just having the highest card to win.

A savvy poker player, or anyone thinking of taking up this kind of challenge, would be wise to consider the mathematical probabilities of the various card combinations. A hand of five cards, from a deck of 52 cards, would mean the number of possible combinations would be:

$$\frac{52 \times 51 \times 50 \times 49 \times 48}{5 \times 4 \times 3 \times 2 \times 1} = 2,598,960$$

In a deck, there are only four royal flushes. The probability of getting a royal flush is therefore $\frac{4}{2,598,960} = \frac{1}{649,740}$ (0.00015%), very small indeed. Gamblers usually like to express the chance of getting a particular hand as 'odds', so in this case it would be odds of 649,739:1. To put it another way, there are 649,739 hands that are *not* royal flushes: a sobering thought. At the other end of the spectrum the overall probability of receiving a five-card hand without *any* matching cards is about 50%.

The low probability of good combinations applies also to one's opponents hands, of course, and like you there is a high probability they may have nothing at all. Other qualities are needed in playing successful poker, since bluffing is involved; in the end, aggressive bidding and the psychological traits of opponents may count for more than the numbers on the cards.

But mathematics enters the game in subtle ways. Professional gamblers use 'card counting' techniques to assess the kinds of cards left in the pack, so they can judge when to raise the stakes and increase their probability of winning. In effect, they use probability theory to increase their chances, and so successful have some of them been that they have been banned from casinos.

The arrival of theory

Although early pioneers had used a working theory of probability, it was not until the 1930s that a satisfactory basis for probability theory was established.

The important figure here was the Russian mathematician Andrei Kolmogorov. In the same way that the Ancient Greeks had highlighted the basic axioms and postulates for geometry, Kolmogorov produced a sound foundation for probability theory. With his work, the *theory* became abstract and removed from its history in games of chance. His contribution established probability as a unified theory to which other mathematicians could contribute, and it became less like a collection of disparate ideas.

With games of chance the probablities can be pinpointed. But significant challenges lie in assigning probabilities to generally stated *events*. How do we answer the question 'What is the probability that it will rain next week?' There are two outcomes: rain or no rain, but, unlike heads or tails, they are not equally likely.

One approach mathematicians have taken is to assign a subjective probability as a 'degree of belief', and to revise this on the basis of evidence.

Such difficulties notwithstanding, probability theory has found a range of wider applications. For scientists, the 'kinetic theory of gases' treats the movement of molecules in probabilistic terms. In daily life, probability is an essential tool in the modern practice of risk assessment, and companies employ actuaries to work out probabilities on which to base their insurance plans. In serious law cases, juries have to know how to interpret the probability of a DNA match, and when they decide guilt or innocence they frequently have to do it 'on the balance of probabilities'.

Eternal optimism

Where there is uncertainty, probability theory is what we need for measuring chance. It may not always be possible to put an exact figure on probability, but knowing how the probabilities combine is knowledge well worth acquiring. Games of chance have, in this sense, spurred mathematicians into evolving a profound theory. We also need, of course, to understand its potential and limitations. No amount of probability theory will guarantee us winning the lottery or scooping the jackpot – but we will get to know the odds, and so we will be better informed, as Monty Hall's contestants were (whether they knew it or not) once he opened a box.

A difficulty, as we have seen, is that mathematical probability can be counterintuitive, and in ordinary life most of us stick with our hunches when faced with decisions to be made. An optimistic sense of possibility persists in the mind of the card player and the lottery punter. We know that mathematics can't *guarantee* riches. But, then again, we just might get lucky …

IS THERE A FORMULA FOR EVERYTHING?

Mathematical recipes and the search for knowledge

*E*veryone likes a shortcut. Looked at simply, mathematical formulae are precisely that – concise recipes, which have been thought out by previous generations so that we do not have to do the work, other than substitute precise values for such 'variables' as x, y or z. But there is more to it than that. The desire to find a formula is a driving force in science and mathematics. For the scientist a formula gives a theory credence and brings knowledge of the world. For the mathematician, it answers a question and establishes a mathematical truth.

In general terms, the word 'formula' suggests a set pattern, a ready-made combination, a more-or-less predictable path. The terms 'formulae' and 'equations' are often used interchangeably, though they are not strictly the same thing. In an equation we are called upon to find the values of a variable that make the left-hand side of the equation equal the right-hand side. A formula is the expression of one variable in terms of one or several others.

Formulae exist in all shapes and sizes. Some are striking for the power they contain; some are amazing, and even beautiful, for the symbols they combine; and some are just plain useful. The world's pre-eminent formula, Einstein's $E = mc^2$, is awe-inspiring in its laconic description of the relationship of energy and mass. But if Einstein's formula is the glamorous star of the repertoire, there is also a crew of others that do vital work.

The backstage crew

The really useful formulae in everyday use are the workhorses we use almost unthinkingly, which enable us to go about our calculations in automatic mode. With a formula like this under our belt we can 'plug and chug' – plug in the value of one variable, do a bit of working out, and come out with the value of the variable we need.

> '*Formulas should be useful. If not they should be astounding, elegant, enlightening, simple, or have some other redeeming value.*'
>
> UNDERWOOD DUDLEY,
> *Mathematics Magazine* (1983)

The two most widely used temperature scales, Fahrenheit (F) and Celsius or centigrade (C), provide a straightforward example. They are named after their 18th-century founders Daniel Fahrenheit and Anders Celsius, and though the former has been superseded in scientific work it is still widely used in weather forecasting, so a formula for conversion is a useful tool.

On the Fahrenheit scale, water freezes at 32 degrees and boils at 212 degrees, so between these two points are 180 degrees of temperature. On the Centigrade scale, water freezes at 0 degrees and boils at 100 degrees so there are 100 degrees separating the two points on this scale. Therefore, each degree in Centigrade corresponds to $\frac{180}{100} = \frac{9}{5}$ degrees on the Fahrenheit scale, so if we want to find the temperature in Fahrenheit we have to multiply the Centigrade degrees by $\frac{9}{5}$. Since the freezing temperature $F = 32$ occurs when $C = 0$, F starts with 32, and the formula is therefore:

$$F = \left(\frac{9}{5} \times C\right) + 32$$

Some formulae are more challenging to construct. We know that the volume of a sphere is dependent on its radius, because if we increase the radius we increase the volume. But what is the exact nature of this dependence? The volume of a cylinder with a circular base is straightforward. It involves multiplying the area of the base circle (obtained by multiplying the square of

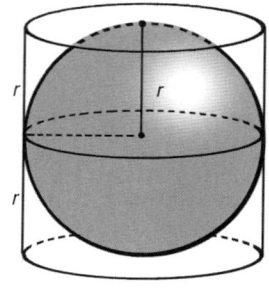

THE VOLUME OF A SPHERE: $\frac{4}{3}\pi r^3$

the radius by pi, that is, πr^2) by the height of the cylinder. But this is not the same as the volume of a sphere.

The formula for the volume of a sphere was discovered by Archimedes, in a moment of sheer inspiration. He stated that it is two-thirds of the volume of a cylinder that encloses it. So the height of the cylinder is the diameter $2r$ and the volume of the sphere is $\frac{2}{3} \times \pi r^2 \times 2r$ making the formula for the volume of a sphere equal to $\frac{4}{3}\pi r^3$.

This brilliant little formula, like so many others, is now a straightforward exercise in the Integral Calculus (see *What Is the Mathematics of the Universe?*).

Basic questions like finding formulae for length, area and volume have often been the source of mathematical progress. The Ancient Greeks were expert on such curves as the circle and ellipse, and in the 17th century the algebra of René Descartes gave us formulae for these curves. While we know formulae for the area and circumference of a circle and the area of an ellipse (equal to πab where a, b are the lengths of its axes), the circumference of an ellipse is not simple, and the attempts to understand its nature have given rise to vast swathes of mathematics on the subject of elliptic integrals and elliptic functions.

The ubiquity of computers in modern life has generated a need for very large numbers of formulae, in the context of 'combinatorics', the branch of mathematics that calculates possible combinations of objects. Three objects, say a, b and c, would yield $3 \times 2 \times 1 = 6$ combinations (*abc, acb, bac, bca, cab, cba*), and a formula is hardly necessary. But if, instead, we were considering ten objects, the number of possible combinations would climb to 3,628,800, and a staggering 9.33×10^{157} for a hundred objects. Here, clearly, a formula would save much toil. This is where 'Stirling's formula', named after the Scottish mathematician James Stirling, comes to the rescue:

$$\sqrt{2\pi n} \times n^n \times e^{-n}$$

Here, n stands for the number of objects. That the formula also includes the mathematical constant π and Euler's constant e (whose value is approximately 2.718...) is a surprise. The presence of π, more usually associated with the circle, and e, associated with growth, is a reminder of the amazing connections that mathematics springs on us, especially since the original problem involves only the multiplication of whole numbers.
Yet the formula is also remarkable for the closeness of its approximation to the true value, and for the case of 100 objects it is only 0.083% out.

A stage beauty

Unlike π and e, the 'imaginary' i cannot be entered into a calculator and evaluated, for it is not a real number (see *Are Imaginary Numbers Truly Imaginary?*). All three symbols frequently grace the mathematical stage, but given their origins they seem unconnected. If we add in 0 and 1 we have all of the famous mathematical constants 0, 1, e, π, i. Boasting all of them, the most 'beautiful' formula in mathematics is surely:

$$e^{i\pi} + 1 = 0$$

It first came to light in the 18th century, and was associated with Leonhard Euler. But how did it come about? We have to go back to the problem of finding 'roots' to see this. The number 1 has two square roots (1 and -1), and we need to conceptualize 1 and -1 at opposite ends of a diameter of a circle. The circle, itself 'beautiful', holds the key to the beautiful formula.

In turn this goes back to the complex numbers, which can be represented by points in the two-dimensional plane (see *Are Imaginary Numbers Truly Imaginary?*). A circle with radius 1 can be thought of as all those complex numbers P which are positioned at a unit distance from the origin at O. It was discovered that $e^{i\theta}$ is a convenient way of representing these complex numbers where θ is the angle that the line OP makes with the horizontal axis.

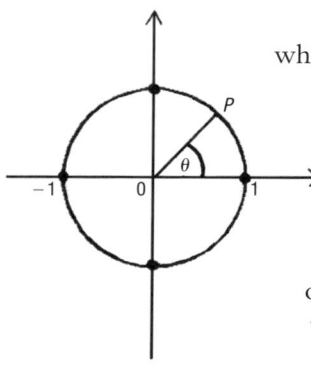

THE GEOMETRY OF THE FORMULA
$e^{i\pi} + 1 = 0$

As the square root -1 is positioned on the circle where the angle is $\theta = 180°$ (which in circular measure is 'π radians'), it means that $-1 = e^{i\pi}$. Finally because of the sum $-1 + 1 = 0$ we have the 'beautiful' formula $e^{i\pi} + 1 = 0$.

A whole series of remarkable formulae are now possible. We have so far considered square roots of 1, but if we consider cube roots, in this case we get three roots of 1 positioned on the circle at angles $0°$, $120°$, and $240°$ (respectively 0, $\frac{2\pi}{3}$, $\frac{4\pi}{3}$ radians), and we gain the formula:

$$e^{\frac{2\pi i}{3}} + e^{\frac{4\pi i}{3}} + 1 = 0$$

Comparable formulae can be obtained by taking fourth roots, fifth roots and so on. This means that our 'most beautiful' formula is only the first one of a very long chain.

A colourful actor

Beautiful formulae exist in geometry, but they can't always be used to give the desired answer. A famous problem in mathematics surfaced in the 1850s: was it possible to colour a geographical map with just four colours, so that adjacent countries would always have different colours? A simple experiment with four adjacent shapes reveals that it cannot be done with three colours. Four colours seemed sufficient for every map looked at, but would it be true for *any* map? And 'any map' to a mathematician did not only mean any known geographical map but any one which could be formed from arbitrary shapes placed on a plane. This became known as the 'four colour conjecture'.

No one could prove this conclusively, but in the 1890s the British mathematician Percy Heawood tackled a related problem. Heawood considered surfaces with holes, such as a 'torus', better known as a doughnut shape. How many colours would guarantee that any map drawn on a doughnut shape could be coloured correctly? Heawood made a spectacular achievement when he produced a formula. For a doughnut shape surface with r holes, Heawood showed that c colours would colour a map where:

$$c = \text{whole part of } \frac{1}{2}(7 + \sqrt{1 + 48r})$$

So, if the surface were a doughnut with one hole, that is $r = 1$, then c would be the whole part of $\frac{1}{2}(7 + \sqrt{49})$, which is 7, and so any map on the torus can be coloured with seven colours. If a surface has three holes, $c = $ whole part of $9.6033 = 9$, and so nine colours would be sufficient to colour any map drawn on this surface.

Heawood's formula was obtained on the strict understanding that the surface had *at least* one hole. In other words, it would be *illegitimate* to substitute $r = 0$ in the formula, though if we did, tantalizingly c would be equal to 4. Unfortunately, a normal, flat map is exactly that, $r = 0$, and, it must also be admitted, there is a dearth of doughnut-shaped maps in the realms of cartography.

In 1976 the problem for maps drawn on a flat surface was finally solved. But no formula was involved, and there is serious doubt whether there is a formula like Heawood's to cover this case. It was shown that any map could be coloured with four colours, provided 1,936 critical map types could be checked. This was done using a computer and the proof of the four colour theorem became one of the first mathematical proofs which relied on a machine.

The supporting cast

We have already raised the fine distinction between formulae and equations. This raises a secondary question, though: is there always a formula for solving an equation?

The quadratic equation in its various forms has been known about since the time of the Babylonians of around 2000 BC. So dealing with the quadratic equation has a long history. To solve a quadratic equation, such as $x^2 - 7x + 10 = 0$, we need to find the value of x that allows the right-hand side to equal 0. The general problem of solving equations can be difficult, but for the specific theory of the *quadratic* equation there is a formula for its solution:

$$x = \frac{-b \pm \sqrt{b^2 - 4ac}}{2a}$$

The letters a, b and c here are any numbers in the general quadratic equation $ax^2 + bx + c = 0$. This well-known formula involves only the basic operations for addition, subtraction, multiplication, division and handling square roots.

Mathematicians of the Renaissance turned their attention to cubic equations, those where the leading term is x^3, and quartic equations, which tackled x^4. In both cases formulae were found. Although in the case of quartic equations it was a long and tedious one, it still involved only the basic operations of handling numbers.

What was more natural now than to seek a formula for the equations of degree 5, the quintic equation, with the leading term x^5? This time, it was easier said than done. For 300 years mathematicians sought it, but none found it. The quest became one of the great unsolved problems of mathematics. Methods that gave *approximate* solutions were possible, but what mathematicians wanted was a formula that gave *exact* solutions. In the 1820s the young Norwegian Niels Abel put an end to (almost) all speculation when he found that *no* formula could exist to cover all quintic equations. For example, a formula will not solve $x^5 - 2x - 1 = 0$. A few valiant souls struggled on in the attempt, but Abel's landmark result settled the question for most.

The glamorous stars

The formulae for solving equations are vital to pure mathematicians, and have resonance for many non-mathematicians from schooldays. But for true glamour in terms of formulae one has to head from mathematics and into the world of science. Indeed, the major contribution that mathematics has made to science has been in focusing attention on formulae.

'Probably no other law of nature has so simply unified any such mass of natural phenomena as has Newton's law of universal gravitation in his Principia Mathematica.'

E.T. BELL,

Men of Mathematics

Thus, Galileo shifted attention away from the question of *why* a stone falls to the earth to the quantitative question of finding a formula for the distance fallen by the stone. On which variable or variables does this depend? Galileo's genius was to discover that the speed of a falling ball did *not* depend on its mass, as was thought by Aristotle. Heavy objects do not fall faster than light ones. Through experimentation, he discovered that the distance d of descent was given by the formula $d = 16x^2$, after x seconds have elapsed. (The factor of 16 is there because the physical units of time measured in seconds, and distance measured in feet have been chosen.) The breakthrough achieved by Galileo was that one simple formula could describe the motion and give answers. Using it we can calculate the distance the stone has fallen at any future time x.

One of the greatest of all formulae is the centrepiece of Isaac Newton's theory of gravitation. We could not put a person on the Moon, engage in space travel, or begin to understand the physical world without it.

Newton theorized that *any* two objects, of masses designated m_1 and m_2, are attracted to each other with a force F that is proportional to the multiplication $m_1 \times m_2$ and divided by the square d^2 (d being the distance between the two masses). The famous formula is:

$$F = G \frac{m_1 m_2}{d^2}$$

111

It is called the 'inverse square law of universal gravitation', and G here is the gravitational constant, whose value depends on the chosen units of measurement. Newton's formula is an 'inverse square law' because we are dividing by d^2, and so the larger the distance between the two objects the weaker the force between them. In the case of planets, m_1 and m_2 are very large and so the forces with which they attract each other will be significant.

The success of Galileo and Newton suggested that workings of the universe were like a mechanical clock and fully determined. It is a powerful idea and one that feeds the motivations of scientists who want to find a formula from their observational work, which might, if they are lucky, become an equally venerable scientific law. Pierre-Simon Laplace, in the late 18th and early 19th centuries, was the great champion of this notion. He declared that to determine the future all that was needed was a vast intellect that could assemble all the instances of the motions of atoms, forces, movement of the planets, and all nature's data, at a single moment – this could then be fed into a formula that would fully determine the future state of the world! Thus, the belief in a deterministic universe was based on the allure of an all-answering formula.

'After relativity, physicists could no longer appeal to a demon who observed the entire universe from outside, but they could still conceive of a supreme mathematician who, as Einstein claimed, neither cheats or plays dice. This mathematician would possess the formula of the universe, which would include a complete description of nature.'

ILYA PRIGOGINE,
Order out of Chaos (1984)

In the 19th century, Michael Faraday knew by experiment that magnetism and electricity were intimately linked. When a bar magnet moves through a coil of wire, a current of electricity flows though the wire, and, the other way around, when a current of electricity passes through a coil of wire a length of iron sitting in the coil is changed into a magnet. The relationship is the basis of the electric motor, but could it be encapsulated in a formula? This is where the genius of the mathematical physicist James Clerk Maxwell comes in: not only did he write down 'Maxwell's Equations', describing this phenomenon, but he invented some new mathematics as well. The four equations making the formula for mixing electricity with magnetism are highly significant.

Formulae describing physical events can be startlingly short, like Boyle's Law ($P = \frac{k}{V}$), which describes the pressure and volume of a gas as inversely related to one another. The iconic formula in physics, Einstein's $E = mc^2$, a product of his theory of relativity, exudes a beguiling simplicity that belies its profound importance. Other formulae, though, depend on several variables, such as the lensmaker's formula, which expresses the focal length of a lens in terms of its refractive index, curvature and thickness.

The magic formula?

In mathematics and science, the existence of a magic formula for everything is illusory. In mathematics, we know there is no algebraic formula which will solve all fifth degree equations, for example. In fact, the 'incompleteness theorems' produced by Kurt Gödel in the 1930s demonstrated that there are statements in formal mathematical systems that are true but undecidable, which rules out an easy formula or a 'machine' to grind out theorems. But aspiration is ever present, and mathematicians continue to look for pattern and formulae. The scientists observe data to see whether it reveals a basic scheme – the precondition for any successful formula. Today's scientists, though less in thrall to determinism, still crave the discovery of a new formula. It just might become scientific law and give authenticity to their theories – and the reward for success: a definite place in history.

WHY ARE THREE DIMENSIONS NOT ENOUGH?

Higher dimensions, monster curves and fractals

*F*or centuries we had just three dimensions and that was enough – they were not questioned. But then physicists asserted that to comprehend the world and the cosmos properly we needed to expand the number of dimensions. More recently, the power of computers to provide vivid graphics has summoned us into a subterranean world where a dimension is not even required to be a whole number. The whole notion of 'dimension' has travelled from the familiar to the elusive.

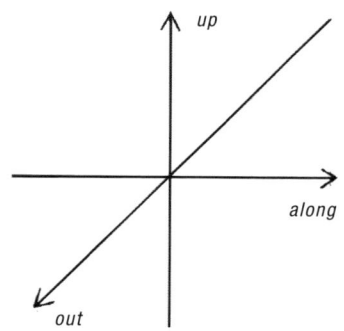

THE THREE TRADITIONAL DIMENSIONS

It is still a commonplace to regard the physical world in which we live as a three-dimensional world. Wherever we are, we can move *along*, *up* or *out* – or a combination of all three. In our culture, the concept of three dimensions retains a strong hold on the language and the imagination: 3D shapes, 3D maps, 3D cinema.

The Ancient Greeks thought of dimension in a hierarchical way. A *point* was the geometrical building block, conceived of as being 'zero-dimensional'. Collections of points made up a line, and so lines were one-dimensional. Lines bunched together created planes – two-dimensional – and finally, planes assembled consecutively created a space that was three-dimensional. Implicit in this hierarchical view is continuous motion: points trace out a line, lines sweep out into a plane, and planes combine to form space.

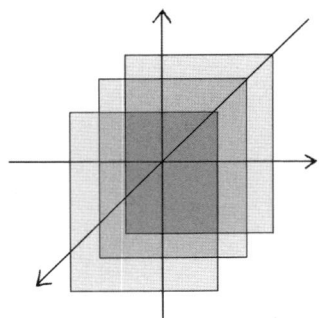

THREE−DIMENSIONAL SPACE FORMED FROM CONSECUTIVE PLANES

Such a fundamental notion as dimension is ripe for exploration by mathematicians and scientists. Aristotle, in the *Physics*, spoke of space being of six dimensions: he thought it quite natural to introduce opposites in motion, hence *along, up* and *out* became the couplings of *back/along, down/up* and *in/out*. But, during the Renaissance, Leonardo da Vinci rejected six dimensions, asserting that the science of painting begins with three-dimensional geometry. Following in the footsteps of the Greek geometers he argued that the point is basic, then comes the line, the plane comes third, and fourth comes space, which is the 'vesture' of planes. Leonardo knew his geometry and thought it appropriate that his students should too.

More than a century later René Descartes revolutionized geometry by changing it into algebra. Using x, y, z axes, subsequently named 'Cartesian axes', a point in space can be pinpointed by assigning numbers to three coordinates (x, y, z). So, for example, (3, 1, 20) represents a point 3 units *along*, 1 unit *up*, and 20 units *out*. This transformation of geometry into algebra meant that geometrical objects could be described and handled by equations. For example $x + y + z = 0$ is the equation of a flat plane and $x^2 + y^2 + z^2 = 1$ is the equation of a sphere with radius 1. These are geometrical objects in the space of three dimensions but the flat plane and the sphere (the surface of a ball) are actually *themselves* of two dimensions. If you lived in and travelled around the surfaces you would think you were living in a space of two dimensions. Using algebra, greater precision could be obtained, and geometry was no longer reliant upon 'seeing' a geometrical figure or shape. The algebraic symbols awakened mathematicians to facets of geometry they might otherwise not have noticed.

The journey into higher dimensions

Initially, higher spatial dimensions, considered unusual, were referred to as 'hypergeometry' (hence 'hyperplane', 'hypersphere'); but in fact it is relatively easy to convey higher dimensions in equations. For four dimensions, we add another variable to the existing three, and for five dimensions we add two more variables. So $x + y + z + w = 0$ is the equation of a hyperplane in four dimensions, and $x + y + z + w + u = 0$ a hyperplane in five dimensions. Likewise $x^2 + y^2 + z^2 + w^2 = 1$ is the equation of a hypersphere in four dimensions and $x^2 + y^2 + z^2 + w^2 + u^2 = 1$ a hypersphere in five (see *What Shape Is the Universe?*).

There is no limit to the number of dimensions we might consider. In proposing the notion of a 'genetic space', the Oxford evolutionary biologist Richard Dawkins set up a computer model, which showed how to navigate through the generations from pre-existing animals to future incarnations. He needed thousands of dimensions, one for each independent gene of the animals he considered.

Some mathematicians took to the hypertheories with alacrity when they were first introduced. Arthur Cayley wrote papers on n dimensions (where n stands for any number) in the 1840s, so there was nothing to stop him thinking of 105 dimensions, or 10,500 dimensions, if he so desired. Nevertheless, deep down, mathematicians were reluctant to abandon the

> '... *you need no more or less than three dimensions to make a knot, a knot that tightens on itself and won't pull apart, and that's what the ultimate particles are – knots in space time. You can't make a knot in two dimensions because there's no over or under* ...'
>
> JOHN UPDIKE (1986)

notion that *ordinary* space, the physical space we live in, remained three-dimensional and that higher dimensions applied only to an abstract mathematical space.

But can we really persist in viewing physical space as three-dimensional? Einstein suggested four dimensions was the correct number and, unlike Newton, who considered time as independent, Einstein thought in terms of a space–time continuum, where time and space affected each other.

Flatland, a two-dimensional world

If we want to appreciate our difficulties in picturing a four-dimensional world, we could not do better than visit Flatland, a conceptual place invented in the 1880s by the schoolteacher Edwin A. Abbott for his social satire *Flatland: A Romance of Many Dimensions*. To construct his world, Abbott collapsed the three space coordinates into just two. In Flatland, people are constrained to live out their lives on a flat surface, without height. They cannot escape, and in particular they cannot attain an *overview* of their world. In Flatland there is no 'above'.

All objects and people in Flatland are reduced to two dimensions, and they can move about in the fictional time of the story but they cannot escape the page. People are represented as shapes, such as triangles, squares, pentagons and other symmetrical figures, and the shapes reflect a rigid class structure. Shapes with fewer sides, such as triangles, are at the lower end of the social hierarchy, and as we progress upwards the shapes gain more sides. Because everyone in Flatland lives in a plane it is hard for different shapes to be distinguished from each other. The inhabitants need to develop subtle ways for recognizing one other.

Take the case of the university professor, a triangle with equal sides. Because all the inhabitants of Flatland live in two dimensions, his students are unable to make out the point of his triangle: all they will see is a line segment AB – that is, unless their view can be modified by light and shade. Women are on the lowest rung in Flatland's social order. They are represented

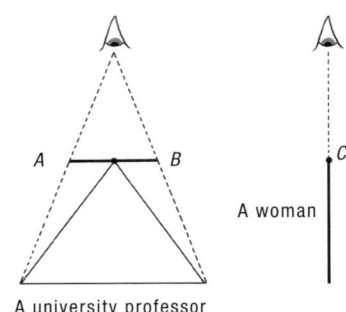

A woman

A university professor

TWO–DIMENSIONAL CHARACTERS OF FLATLAND:
A PROFESSOR AND A WOMAN

as 'straight lines', and there is no way a woman can gain status, for she cannot advance in the social hierarchy to become a triangle. But women do have one unique quality. Being lines, they can become almost invisible: head on, all that can be perceived is a single point.

Social satire apart, the perceptual and conceptual difficulties experienced by the Flatlanders are identical to the ones we experience in trying to visualize a four-dimensional world, let alone one of higher dimensions. How do we 'see' a hyperplane or hypersphere?

Cantor's surprise

If geometers of the 1870s thought they understood the concept of dimension, they were about to be surprised. Dimension had been grasped intuitively, but there was nothing like a watertight definition of it. Then Georg Cantor, working at the University of Halle in Germany, made a startling discovery. Traditionally, a square was seen to acquire its two-dimensional nature from the idea that it consisted of lines of unit length piled one on top of the other. If a one-dimensional line of unit length contains a number of individual points that can be marked, then a square, consisting of *numerous* lines, was thus supposed to have many more points within it.

Cantor contradicted this notion. He set up a one-to-one correspondence between the points on a line of unit length and the points in a unit square. In other words, he claimed that every point on the line had a corresponding point within the square, and vice versa.

How did Cantor do it? He first expressed a point on the line as a decimal – for example, the point might be 0.19762543… He then created a corresponding point in the square with coordinates (x, y) by means of *alternately allocating* the digits to x

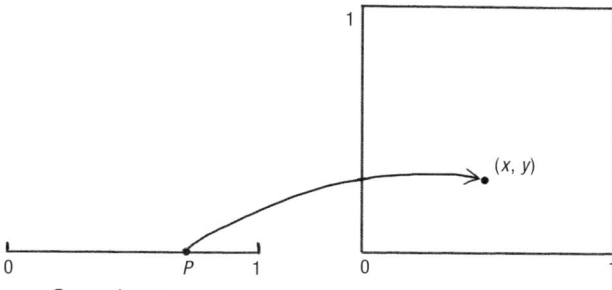

CANTOR'S ONE–TO–ONE CORRESPONDENCE OF LINE AND SQUARE

and y, so that here we would have $x = 0.1724\ldots$ and $y = 0.9653\ldots$ In reverse, if we are given a point in the square (x, y), we can interleave the digits of x and y to form one decimal expression corresponding to a point on the line.

What Cantor had done flew in the face of intuition, even his own. He wrote to Richard Dedekind in 1877, remarking: 'I see it, but I don't believe it'. Until then, geometers had treated one-dimensional and two-dimensional objects as *radically* different, but Cantor's bold step concluded that they possessed, via this one-to-one correspondence, the *same* number of points. He could even set up a one-to-one correspondence between a line and a three-dimensional cube (using a similar method and taking every third value), and beyond that into higher-dimensional cubes.

For mathematicians there was one property missing from the one-to-one correspondence Cantor had devised. It was not a *continuous* correspondence, that is, Cantor could not guarantee that points *near* a selected point on the line all transformed into points on the unit square near the corresponding point. Twenty-five years later, it was shown by the Dutch philosopher and mathematician Luitzen Brouwer that it was not possible to transform objects of different dimension one into the other by a continuous one-to-one transformation.

Curves and 'monster' curves

Entangled with the problems of defining a dimension, and the numbers of them, was an enduring question: what is a curve?

The Ancient Greeks had dealt with circles, ellipses, hyperbolas and spirals and other special curves. But what about curves that did not fit into the classical mould? In the 1880s, Camille Jordan tentatively defined a curve in terms of a continuously moving point.

The advantage of this definition is that quite complicated images, such as unbroken 'fingerprints', can be regarded as curves. The trouble was that it admitted too much into the fold, for by this definition a *solid* square could also be a curve. How could a 'one-dimensional' curve be a solid 'two-dimensional' object? This was unreasonable. Attempting alternative definitions, mathematicians began producing 'curves' with the strangest properties, unintended consequences that were completely outside the class of curves normally encountered. They were, appropriately, dubbed 'monster' curves. These investigations into curves also generated a whole range of different, but interrelated, concepts of dimension.

One of the monster curves produced at the turn of the 19th century was the Koch curve, named after the Swedish mathematician Helge von Koch. It is a 'monster' in terms of its properties but not in terms of its appearance – for it is the much reproduced snowflake curve. To generate it, one starts with an equilateral triangle (defined as 'stage 0'), introduces a triangular-shaped 'kink' in each side, and then continues repeating the process, a succession of stages n with the snowflake curve being the limit of the stages. The curve is thus generated by repetition. It is not possible to draw the *actual* snowflake curve, but it can be represented at one of its stages.

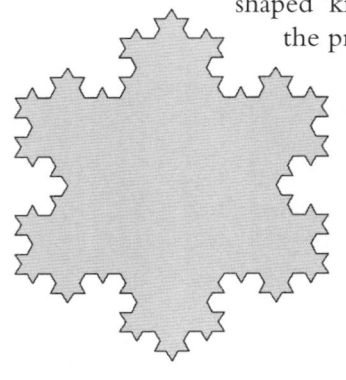

A STAGE OF THE SNOWFLAKE CURVE

As more kinks are added the boundary of the snowflake increases, and in fact the length of the ultimate snowflake curve is infinite. Because the shape always stays within a circle it has a finite area, so it is a curve with a *finite* area and an *infinite* bounding curve.

But what is the dimension of the snowflake curve? How can we measure it and which of the many definitions of dimension available should we use? This is where the ideas of the German mathematician Felix Hausdorff prove useful. The 'Hausdorff dimension', as it is known, conveniently coincides with 'ordinary' nomenclature for normal shapes, so that a line is 1, the dimension of a square is 2, and a cube 3. This is because the Hausdorff dimension (d) depends on measuring length and area. If a square is scaled up on each of its sides by a factor of 3, the new area is nine times its previous value. As $9 = 3^2$, the Hausdorff dimension of a solid two-dimensional square is the power $d = 2$. This is, of course, the value we expect for a square. The power is the key to Hausdorff dimension.

The generating element of the snowflake curve is the segment of a line and if we scale this up by a factor of 3 and introduce the kink we have a line portion four times longer than the original. So we have to figure d where $4 = 3^d$. The value of d must be between 1 and 2 because 4 is between 3^1 and 3^2 and in fact $d = 1.29224\ldots$

This is something new. We have a curve whose dimension is not a whole number – it has fractional dimension. We have, in fact, entered the world of what, over 50 years later, would include curves known as 'fractals'.

Fractals and the Mandelbrot set

When we take a magnifying glass to the Koch curve we find something curious. It looks the same under the magnifying glass as it does without it. And if we zoom in still further we find the same thing. The Koch curve has the property of being self-similar, a consequence of it being generated by repetition. This is not something that occurs with ordinary curves, such as the circle. If we zoom in on part of the circle's circumference, we see a curved line that becomes straighter and straighter the closer we get – we do not see lots of small circles.

We can thank the 'father of fractals', mathematician Benoît Mandelbrot, for bringing the science of fractals into modern consciousness. A fractal is a sort of fractured or broken curve or shape incorporating self-similarity and the measurement of dimension by Hausdorff dimension. The Koch curve was built up from straight lines which have ordinary dimension of 1 but the Koch curve has Hausdorff dimension, $d = 1.29224\ldots$ and it is typical of a fractal that this dimension should exceed its ordinary dimension.

The best-known fractal of all is the Mandelbrot set. It was discovered by Mandelbrot in the computing laboratory in 1980, and it is even stranger than the Koch curve. When he saw it first appearing on his line printer Mandelbrot was astonished by its intricacy. It was quite unexpected, and because it is found from a basic situation it has become emblematic of all fractals. Mandelbrot discovered the set, not by generating kinks by repetition, as for the Koch curve, but by the repetition of a basic algebraic formula.

The Mandelbrot set too is self-similar; if we were to zoom into a part of it we would see millions of miniature Mandelbrot sets. And if we zoomed into each of these, we would see millions more. But more surprises are in store when we measure its dimension. The boundary of the Mandelbrot set is not at all like a square but it does have Hausdorff dimension equal to 2, the same Hausdorff dimension of the square, once again highlighting the bizarre nature of dimension itself.

Once fractals became widely known, they expanded the range of mathematical and computer modelling. One can't attempt to model cloud formations using the objects

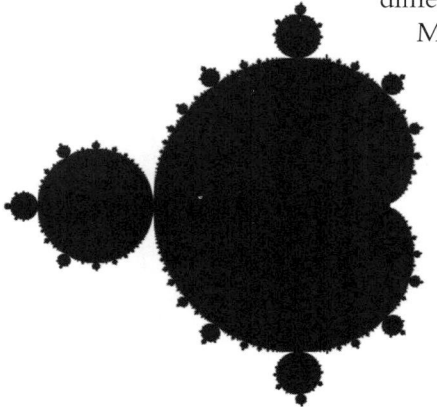

THE MANDELBROT SET

of Euclidean geometry. Clouds are hardly comparable to circles or ellipses; but they *are* like fractals.

Fractals threw new light, too, on practical problems, for example in geographical mapping. We would think that the measurement of the length of a coastline would be straightforward and uncontentious. But the answer you get depends on the unit you use for measuring it. If you fit a kilometre unit around the coastline you get one answer but if you fit a metre unit you get a much longer coastline. This is because the smaller unit can be more closely fitted to the nooks and crannies of the coastline than the larger unit. If the measuring rod were equal to the diameter of an atom, the length of the coastline would be very long indeed. At each stage of the snowflake generation the kinks get shorter but the total length gets longer.

Our multi-dimensional world

From the Ancient Greeks who counted one, two and three dimensions, mathematicians have revealed scenarios and concepts where many more dimensions – and fractional dimensions – come into play. 'Pure' mathematicians can now have as many dimensions as they please, and practical scientists too have felt a need to escape from the world which imposes three dimensions on them. Furthermore, in the 21st century, theoretical physicists in universities across the world have moved on from Einstein's space–time continuum of four dimensions to attempt to locate an all-embracing physical theory. Based on a conception of the universe as consisting of 11 dimensions, structured by tiny 'vibrating strings', these researches have manifested themselves most exotically in string theory, which is, to date, as unprovable as it is tantalizing. In mathematics, and beyond, the concept of dimension continues to exert a powerful grip on the investigative mind.

CAN A BUTTERFLY'S WINGS REALLY CAUSE A HURRICANE?

Chaos theory, weather equations and strange attractors

A swallow-tail, a most beautiful species of butterfly, sits on a bush in the upper reaches of the Amazon where no human has ever visited. It flaps its wings and resumes its basking in the Brazilian sunshine. It cannot know that this brief movement will result in a hurricane thousands of miles away to the north. This vivid image, a bold piece of advertising for 'chaos theory', captured the public's imagination. It contributed to creating a new field of study, one that, along with relativity and quantum mechanics, has been hailed as one of the 20th-century's major advances in science and mathematics.

Let's conduct an experiment. We wind up a clock and set it going at 7.00 a.m. in the morning, and at the same time set another at 7.01 a.m. beside it. During the day the second clock remains, unsurprisingly, one minute ahead of the first. This is the way we think of the mechanical universe, working like clockwork. But not all systems work in the same way. The development of identical twins, for example, is far less predictable. At birth, there may be few identifiable differences in appearance and behaviour. For the first few years their lives will likely progress in a similar way, but gradually individual personalities will emerge and their life paths will begin to diverge. By adulthood, they may be living quite distinct kinds of lives.

In a rudimentary way, we have summarized 'chaos theory', which holds that, over time, systems follow an evolutionary path and diverge as a consequence of small changes or differences in original circumstances. In nature, the theory of evolution suggests the grandest narrative of chaos theory in our world. But, mathematically speaking, the popularization of chaos theory is much nearer to our own time than it is to Darwin's.

Coffee breaks and butterflies

In the 1960s, the US mathematician and meteorologist Edward Lorenz was working on the Navier–Stokes equations, used to predict the weather. These equations express the velocities of weather flows in terms of the x, y, z coordinates of points in three dimensions, so that solving them amounts to finding the paths traced out by these points. Lorenz had to settle for a computer-based solution, because there was no way a solution in exact terms could be found. He set the computer the task of trundling through the millions of calculations required by the numerical techniques, and he began by inputting initial values for x, y, z, before heading off for a coffee break. On his return he found his computer churning out the pattern of an evolutionary path – but there was a problem. The path was at variance with one he had earlier obtained from a similar experiment using the same equations.

After some thought and a moment of revelation, he realized that for his second experiment he had rounded his earlier initial values x, y, z. Where he might have earlier used a value of 5.8904533 for x (say) he had subsequently rounded it to slightly fewer decimal places, at 5.89045, before going for his coffee break. In effect the computer had calculated two evolutionary paths based on the minutely differing initial values.

'Of course the word chaos is used in rather a vague sense by a lot of writers, but in physics it means a particular phenomenon, namely that in a nonlinear system the outcome is often indefinitely, arbitrarily sensitive to tiny changes in the initial condition.'

MURRAY GELL-MANN,
NOBEL LAUREATE IN PHYSICS (1969)

That tiny mathematical difference was one that Lorenz elaborated into the indelible analogy of chaos theory in his 1972 lecture 'Predictability: Does the Flap of a Butterfly's Wings in Brazil Set off a Tornado in Texas?' The key idea he sought to express was *sensitivity*. Our aforementioned mechanical clocks, set at 7.00 a.m. and 7.01 a.m., do not possess this sensitivity because, as time progresses, they keep in step. (Of course, we know that in the *real* world mechanical failure must set in at some point, and one clock will last longer than another before needing to be wound up, but we can set those issues aside: the point here is the clockwork, mechanical principle.) But in the example of the twins, small differences at birth were magnified as time went on. 'Sensitivity due to initial conditions', to give it its full terminology, suggests that small differences at time zero can lead to large ones later. In the field of meteorology, Lorenz drew attention to the fact that some weather equations could predict wildly different weather when the initial values were very close to each other.

Poincaré and the origins of chaos

In fact, the 'butterfly effect' was known well before Lorenz gave his lecture. Henri Poincaré, over 60 years previously, had known about it and had even had weather forecasting in mind when he wrote: 'It may happen that slight differences in the initial conditions produce very great differences in the final phenomena ... a slight error in the former would make an enormous error in the latter.' For Poincaré, though, unaided by the computational abilities of computers, these 'slight differences' appeared to confound rather than augment the ability to predict: 'Prediction becomes impossible.'

Poincaré made these observations as he pursued his astronomical studies on the movement of three bodies. The problem of describing mathematically the motion of two bodies moving in tandem, like the Moon circling the Earth, is straightforward. The challenge for Poincaré was doing the same for *three* bodies, which might be the dynamics of the Moon, Earth and Sun or, at the sub–atomic level, the dynamics of an atom of helium, with its nucleus and two orbiting electrons.

In the dynamics he explored, Poincaré had, in effect, found 'chaos'. Over time, the motion of the three bodies diverged a little from their previous paths. His work suggested a new mathematical phenomenon, which others soon pursued further. In the United States, George Birkhoff took the study of dynamical systems into topology, the pure-mathematical geometry concerned with the ways in which points are connected and whether one surface can be transformed into another (see *What Shape Is the Universe?*).

But chaos was not a wholly abstract study. Poincaré may have encountered it in his astronomical investigations, but chaos exhibited its effects in practical subjects too. An electric diode modelled by equations showed the same sensitivity to initial values. The new research area of chaos theory came into being, linking apparently disparate subjects such as mechanics, radio oscillations, control theory and pure mathematics.

> *'Why have the meteorologists such difficulty in predicting the weather with any certainty? … One tenth of a degree more or less at any point, and the cyclone bursts here and not there, and spreads its ravages over countries it would have spared.'*
>
> HENRI POINCARÉ

Strange attractors

Poincaré used a new sort of diagram in an abstract space of points to describe the different phases of motion of a dynamical system. For him this *phase-space* was a vital window on the dynamics, so let's look at the ordinary pendulum and see how this works out.

The simple pendulum is a rigid rod attached to a fixed point at one end and with a bob at the other, and it swings to and fro. The variables which describe its motion are x, the angle made with the vertical, and y, the angular velocity of the bob. So, for example, at the start the y value will be 0. If we start the pendulum off from some position S and let it go it will gradually lose energy and come to rest, so in the long run $x = 0$ and $y = 0$. Its motion can be described by the paths in two dimensions joining up the points

(x, y). Whatever the initial starting point S of the bob, it is totally predictable that the path will spiral down to the origin $(0, 0)$ as the pendulum slows down and its swing decreases.

The 'attractor' in this case is simply the place where it comes to rest – the point $(0, 0)$ at the origin in two-dimensional space, where displacement is zero and velocity is zero. For any starting position of the pendulum, this point *attracts* any spiral that describes the motion of the pendulum.

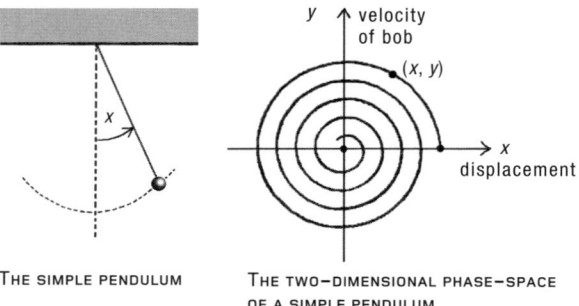

THE SIMPLE PENDULUM

THE TWO-DIMENSIONAL PHASE-SPACE
OF A SIMPLE PENDULUM

The pendulum in a grandfather clock is slightly different, because it swings to and fro without stopping. The path that describes its motion does not spiral into the origin but moves in a circle. If its motion is slightly disturbed – say, the clock receives a bump – the path that describes this motion will be attracted back to the circle.

Neither type of pendulum fits the mathematical sense of 'strange'. That adjective applies when the attractor is neither a point nor a circle but rather, in the words of pioneer David Ruelle, a 'weird set of points' which attract the paths as they thread their way around space. Ruelle described, in decidedly unmathematical language, 'clouds of points', sometimes evoking 'galaxies or fireworks'. In his eyes, 'strange attractors' was a term 'well suited to these astonishing objects, of which we understand so little'. The designation caught on and quickly captured the scientific and popular imagination.

In the case of Lorenz's meteorological investigations, the solution paths that were determined by his weather equations wound around such a set of 'weird' points in three-dimensional space. The visual image produced – the Lorenz attractor – became emblematic of this new science of chaos.

But one outstanding problem remained: did the Lorenz attractor actually *exist*? Was it an artificial construct generated by numerical methods rather than an intrinsic property of the Lorenz equations themselves? The question was answered thirty years after Lorenz's discovery. It was confirmed: the Lorenz strange attractor existed and it was a fractal (see *Why Are Three Dimensions Not Enough?*). Once it was established, the next stage was to measure its fractal dimension, which was found to be around 2.05.

Populations and the chaos principle

Chaos concerns developments over a period of time. In the Lorenz weather equations, time flows continuously. But chaos theory was found in simpler equations too, where time is measured by discrete values 0, 1, 2, 3, etc., such as the year 0, year 1, and so on.

THE LORENZ ATTRACTOR

Population growth had long been theorized in terms of equations, but no one had thus far looked at them from a chaos perspective.

The most straightforward population model gives the population at the beginning of a given year as a multiple of the population at the beginning of the previous year. If the multiple, or growth rate is $r = 2$ (say), and the population at the beginning of a year is half a million, the population at the beginning of the next year will be twice that, one million; at the beginning of the next year after that, it will be two million. In this step-by-step mechanism we count the time in $n = 0, 1, 2, 3$, etc. years. In effect we are, in mathematical notation, finding the population P_{n+1} in one year from the population P_n in the previous year by the basic equation

$$P_{n+1} = r \times P_n$$

applying it one step at a time.

However, this type of equation yields an unrealistic conclusion. Because the population grows proportionately (which defines the equation as *linear*), the population goes on increasing

129

forever. Censuses and history show that this is simply unrealistic. A better equation is needed.

A refinement is the equation

$$P_{n+1} = (r - sP_n) \times P_n$$

where the new growth rate $(r - sP_n)$ depends on the current population size, and reduces as the population increases. If, as before, the initial population is half a million, $r = 2$ and we take $s = 0.04$ (say), from this equation we can compute the population at the beginning of the next year as 990,000, and the year after that 1,940,706. Already we see that these are less than the values forecasted by the basic model of one million and two million.

With this equation it is much easier to work with a more streamlined version, the (famous) logistic form of it:

$$x_{n+1} = r(1 - x_n) x_n$$

Here the variable x_n is measured on a scale from 0 to 1, but we can easily recover the actual population size by multiplying our answers by the ratio of r to s, here $\frac{2}{0.04} = 50$, if we need to. But here we are now more interested in the equation itself. This takes us into the realms of chaos.

The multiplication of x_n and $1 - x_n$ makes the equation *non-linear*, and this is why, with this equation, we are now dealing with the chaos concept. Mathematically, chaos is a property of non-linear equations.

If we take $r = 2$, as before, we find that *whatever* the initial value in year zero, that is, x_0, the population values year by year will gravitate to a consistent long-term value of 0.5. (If $x_0 = 0.6$, for instance, the step-by-step answers will be $x_0 = 0.6$, $x_1 = 0.480$, $x_2 = 0.499$ and so on, and this quickly settles down to the value 0.5.) In the language of chaos, the unique value 0.5 'attracts' all sequences of values to it: it is the attractor.

If we were to push the value of r in the logistic equation a little beyond 3, we would find that there is not *one* long-term value but *two*, so that the population values oscillate between them year-by-year. This is because of the instability of the (so-called!) 'fixed point' with this r value. For instance, if $r = 3.2$, the two attracting points are approximately 0.513 and 0.799, and – again assuming a basic value of $x_0 = 0.6$ – the logistic equation would produce values in successive years of 0.768, 0.570, 0.784, 0.541, 0.795, etc., flitting between the two attracting points.

By increasing the value of r still further we find the number of attracting points are doubled and we get 4, 8, 16, 32 and more attractors for increasing values of r. When the attracting values become so numerous, though, prediction becomes difficult. When we get to $r = 3.57$ and beyond, as far as 4, there is no longer any discernible pattern, and prediction is impossible. Turning up the values of r resembles the turning on of a dripping tap. At first the drips are regular, identifiable, but as the tap is turned up the drips develop into an indistinguishable stream.

With high values of r the butterfly effect truly comes into being, whereby very close values of the original population can still develop in markedly different ways as time progresses. If we take $r = 4$ and apply it to the close-but-not-identical initial population values $x_0 = 0.600$ and $x_0 = 0.610$ we find that even after the first few values, by year six (x_6), we have arrived at population values 0.025 and 0.529 respectively.

The shape of the 'chaotic', divergent pattern can be represented in a 'bifurcation diagram'. It shows how the separate attracting points are born with the increasing values of r. For the starting value x_0, measured on the vertical axis (again, $x_0 = 0.6$) it shows where the fixed point splits into two prongs of a pitchfork shape for different values of r. So there is

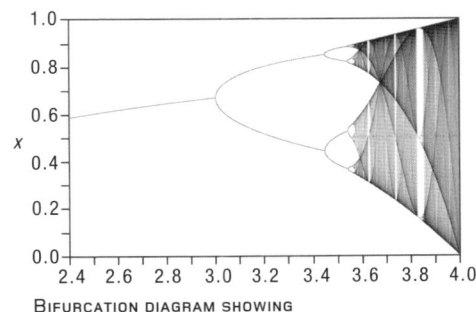

BIFURCATION DIAGRAM SHOWING
CHAOTIC DIVERGENCE

a *unique* limiting value for *r* up to 3, but past 3 the number of attractors increase to two, four, eight … and on to chaos!

An explosion of chaos

Since chaos came to light there has been a vast outpouring of published work, and it has since grown into a wide-ranging and variegated theory, which is still rapidly developing. Its application in mathematical modelling has also prospered. Today, scrutinizing the non-linear equations which govern the motion of ships in heavy seas, engineers can pattern roll and sway and so take precautions against the possibility of ships capsizing. From the movements of financial markets to the rhythms of the heart, chaos theory and non-linear equations have made their mark.

Impressive though these applications are, and powerful though the grip of chaos on the imagination may be, there are limits. The chaos principle, dependent as it is on the sensitivity of initial conditions, suggests that forecasting *some* physical processes is theoretically impossible. Long-range weather forecasting falls into this category, since it is predicated on knowing the weather conditions *exactly* at some point in time, about which we cannot be sure. The inevitable inaccuracies would be enough to make the forecast wrong.

To put it another way, it is entirely up to that swallow-tail butterfly in Brazil whether it chooses to flaps its wings – or not.

CAN WE CREATE AN UNBREAKABLE CODE?

Ciphers, the Enigma machine and quantum computers

When artist and entrepreneur Samuel Morse sent his first coded message 'What hath God wrought' from Washington to Baltimore in 1843, he had invented a brilliant code. It was to be used the world over, wherever the technology of the telegraph existed. It was, of course, a code intended to be intelligible, a concise and practical means of transmitting a message. But historically there has been a whole other world — of secret codes and ciphers — to fascinate us. From early warfare to 21st-century computer systems, numbers have played a vital role in codemaking and codebreaking.

It is a truism to say that communication lies at the centre of the human experience, but, social animals that we are, a world without communication is unimaginable. Ordinary languages are codes — conventions, with rules, that determine who does, and who does not, understand them. But by the term 'code' we generally think of something else, something more purposeful and deliberately divorced from ordinary language, often something more targeted and discreet, even secret. In that sense a code becomes a way of communicating between specific groups and individuals (and, in the computer age, machines), where prying eyes are unwelcome.

From cipher to code

A historically popular method of code creation has been using a cipher. This means changing around the letters in the message according to some system or substituting another set of symbols for the letters. The message, or 'plaintext', is encrypted into a 'ciphertext', which is then decrypted by the receiver.

An early example of a cipher was the 'three-shift system' used by Julius Caesar. To encrypt a message each letter in the plaintext is substituted by a letter that lies three letters further on in the alphabet. So the letter a is replaced by *d*, b is replaced by *e*, and so on. When we come to x and we have exhausted the alphabet, we substitute *a* as if the alphabet were starting again. The encryption of the plaintext 'enemy in next town', for example, would come out as *hqhpblqqhawwrzq* (leaving out spaces).

a	b	c	d	e	f	g	h	i	j	k	l	m	n	o	p	q	r	s	t	u	v	w	x	y	z
d	e	f	g	h	i	j	k	l	m	n	o	p	q	r	s	t	u	v	w	x	y	z	a	b	c

THE THREE-SHIFT CIPHER SYSTEM

As the simple-minded 'Caesar-shift' cipher became better known, and thus insecure, new and more intricate methods of encryption needed to be devised. In more secure ciphers, letters are manipulated in a more ingenious way than just shifting the alphabet along.

When encryption involves substituting letters for whole words, or groups of words, in technical terms we speak of a 'code'. As distinct from a cipher, it requires a 'dictionary' of codewords, which both sender and receiver would possess. In common usage the distinction is often blurred.

Codes, keys and a Greek historian

As advanced methods of codemaking proliferated, the services of specialized codebreakers became a necessity. The typical codebreaker needs to be a combination of mathematician, statistician, linguist and puzzle solver. He, or she, needs to be imaginative, methodical, tenacious and have the ability to winkle out the smallest clue and turn it to account in decrypting a message. Codebreakers need to understand the tricks of the encrypters and they need to keep abreast of coding trends.

Codebreakers know that for a code to work it needs rules, and therein lies a potential vulnerability – there is the potential for decoding, if the rules can be discovered. If letters were substituted for other letters *at random*, a message could be encoded but it could not be decoded by the intended recipient: a waste of time. Often the rules translate into a 'key' – a governing principle or layout that unlocks the workings of a code. It is worth looking at the workings of keys in some other historically significant codes.

During the First World War, codemakers were fond of the 'Polybius 5 by 5 square', named after a Greek historian Polybius, who used it as a way of encrypting letters by *pairs* of numbers. In the square, each letter corresponds to a pair of numbers, one from the row and one from the column. For example, E is in row 1 and column 5, and so E corresponds to 15. Using the Polybius square the word 'enemy' would be encrypted as 1533153254. Obviously anyone who had the actual Polybius square could decrypt the message, so it provided the key to the system.

	1	2	3	4	5
1	A	B	C	D	E
2	F	G	H	I/J	K
3	L	M	N	O	P
4	Q	R	S	T	U
5	V	W	X	Y	Z

THE POLYBIUS 5 BY 5 SQUARE

In 1918, Colonel Fritz Nebel devised a method in which the Polybius square was written with the letters *A, D, F, G, X* instead of numbers 1, 2, 3, 4, 5. These particular letters have different Morse code sounds and so they were less likely to be confused with others when transmitted. With this 'security upgrade' of the Polybius square, the alphabet was also randomly arranged in the table.

The message 'enemy in next town' might then be encrypted as the mystifying: *XDGAXDFXDXDGGAGAXDA DFAFADAAFGA*. As a further embellishment, a *secondary* key might be introduced to scramble the encrypted message before it was sent. But codes with keys of this type have inbuilt vulnerabilities and can be broken almost immediately.

	A	*D*	*F*	*G*	*X*
A	B	X	W	K	C
D	O	Z	A	I/J	Y
F	T	F	U	P	M
G	N	L	Q	H	R
X	S	E	V	D	G

Fritz Nebel's improvement of the Polybius square

Machines and mathematicians

Encrypting a message by hand using a scheme such as the Polybius square was a long-winded process, and errors could creep in, especially when done in conditions of war and battle. It is a mechanical process, so why not actually use a machine? 'Enigma' (derived from the Greek word for puzzle) was a German device, which first saw the light of day in 1923 when it was marketed not for military use but as a means for transmitting commercially sensitive material. It went on to epitomize the successful Bletchley Park codebreaking effort in the Second World War, when its secrets were finally uncovered by cryptographers.

The Enigma machine manipulated input using a combination of mechanical components and electrical circuitry – an electro-mechanical device. It was based on three discs (rotors) arranged on a single spindle and each marked by the 26 letters of the alphabet. The electrical contacts on one side of each rotor were hardwired to the contacts on the other side, and each rotor was wired differently.

On the machine there was an alphabetic keypad and a set of alphabetic characters, which could be illuminated. When the A key was pressed a different letter, say P, lit up; and the first rotor rotated one step. After 26 presses the first rotor had gone through one revolution and the second rotor turned one step;

26 presses later, the third rotor would come into action. Each time a key was hit, a different circuit in the machine was created. If the sender had to type the letter A later on, a different encoding would ensue, so that P would not necessarily be lit up. It was this 'polyalphabetic' substitution that gave the Enigma machine its complexity and its power against codebreakers. By calculation, on the basic model the electric circuits would not be repeated until after $26 \times 26 \times 26 = 17,576$ key presses: a forbidding array of possibilities.

Ideally an encryption machine should present no characteristics that a codebreaker might exploit. This is where Enigma became vulnerable. In addition to the *cyclic* nature of the encryption, caused by the rotors, it had two other weaknesses:

- A letter could not be encrypted as itself, so the encryption of A had to be a letter other than A.
- Enigma was reversible. If A was encrypted as P, then pressing the P key (with the discs in the same position) resulted in A. This helped the receiver but it also helped the codebreaker.

The sender gave the machine an initial setting with three letters uppermost on the rotors – it might be RSM for example. Let's suppose the sender wishes to send the plaintext message 'Enemy in next town', and the receiver gets the scrambled ciphertext PBNTAMJGLQWDLFZ. To decipher it the receiver sets his identical Enigma machine to RSM and works in reverse, typing in the ciphertext to retrieve the original plain text. It is clearly vital that the receiver knows the sender's starting position key, and for the intended receiver this would be set out in a codebook listing the daily keys.

If 17,576 starting positions was not thought to provide enough security, the odds against decipherment were soon increased – and dramatically. This is where the mathematics comes in. With three rotors *fixed*, the starting positions could be increased, allowing a *choice* of $6 = 3 \times 2 \times 1$ possible placings of the rotors on the spindle. The increase was thus to $6 \times 17,576 = 105,456$ effective starting positions. When the possibility of choosing three rotors from a set of five was introduced, there was a further increase to $10 \times 105,456 = 1,054,560$ starting positions.

This was all before the introduction of the plugboard which fitted on the front of the Enigma machine. This was a means of *swapping* some letters of the alphabet in pairs by interchange leads on the plugboard. For each starting position there were 3,453,450 ways of connecting up just three pairs of alphabetic symbols, but more than a staggering 150 million million ways of connecting up ten pairs of letters for *each* starting position!

Given the scale of these encrypting possibilities, it was therefore essential for the code breakers to use a 'crib', a starting word that might give a clue to the key and which could enable false trails to be discarded. A message might start with 'Top Secret' or 'Heil Hitler' for example.

The communications between members of the German high command was encrypted by the Lorenz machine, an encrypter used in conjunction with teleprinter traffic. It had no less than 12 rotors available. To decrypt these messages, *Colossus*, the world's first programmable computer was introduced, designed by British telecommunications engineer Tommy Flowers.

Public key encryption

If the existence of a key, mechanical or otherwise, created the security vulnerability for codes and ciphers, what might be the solution? The answer, when it came, was a big surprise – make it public! Well, not quite, for there would be some vital information known only to the receiver.

'Public key encryption', as it became known, begins by distributing chosen numbers. We can work through this with an example, so that, say, numbers 3 and 55 are made available to all and sundry. They may be broadcast on the radio or made available at the local library. Let's suppose our familiar secret message 'enemy in next town' is coded '18' in a codebook also widely available. If an eavesdropper knew that the message was '18' they could simply refer to the codebook.

However, using the two numbers 3 and 55 we can encrypt message '18' and transmit it, so '18' would be kept secret from the eavesdropper. This is where some very old number theory comes into play, of the kind that was once regarded as possessing no possible earthly applications. Using the number 3, we first calculate

$$18^3 = 18 \times 18 \times 18 = 5{,}832$$

Now we divide this number by 55 and work out the remainder left over after division. Because $5{,}832 = 106 \times 55 + 2$, the remainder is 2. We now take that value 2 as the encryption of '18' and we transmit this, remembering that *anyone* can encrypt a message, because numbers for 3 and 55 are in the public domain.

So the receiver now reads 2 and has to decrypt it. The receiver knows something about the number 55 that is *not* publicly available – that it is the product of two prime numbers, $p = 5$ and $q = 11$. This is where a theorem discovered by Pierre de Fermat coupled with the Chinese remainder theorem comes into play, a potent combination of pure number-theory magic.

> *'If all the personal computers in the world, about 260 million computers were put to work on a single P[retty]G[ood]P[rivacy]-encrypted message, it would still take an estimated 12 million times the age of the universe, on average, to break a single message.'*
> WILLIAM CROWELL,
> DEPUTY DIRECTOR OF THE US NATIONAL SECURITY AGENCY,
> DESCRIBING THE PGP PUBLIC KEY ENCRYPTION SYSTEM IN 1997

Using the other public number 3, this guarantees that there is a value of x which has the property that $3 \times x$ leaves a remainder 1 when it is divided by $(p - 1) \times (q - 1)$. So the receiver needs to figure out this value in the case of the primes $p = 5$, $q = 11$ where $(5 - 1) \times (11 - 1) = 4 \times 10 = 40$.

We know that such a number exists, but we still have to figure it out. In our case, with such small numbers, we see that $3 \times 27 = 81$ and by dividing 81 by 40 there is a remainder of 1, and this means that the value of x is 27. This is the number we need to decrypt the number 2 we have received.

To do this we work out $2^{27} = 134{,}217{,}778$. When we divide this by 55 we get 2,440,322 and remainder 18 because:

$$134{,}217{,}728 = 2{,}440{,}322 \times 55 + 18$$

The remainder is the original message '18', now decrypted.

The security of public key encryption depends on the receiver knowing a secret element, in our example that $55 = 5 \times 11$. With such a small number as 55 it is easy to figure out the prime factors 5 and 11. But with a number of 1000 digits, say, as would be the normal practice, it is just about impossible. It is much easier to multiply two prime numbers together than to perform the inverse operation of deriving prime factors of a given number. Given any number, it is possible to answer the question 'Is this number a prime?' by an efficient computer algorithm. But *finding* the prime factors is a totally different problem, and no efficient computer algorithm is known for it (see *Why Are Primes the Atoms of Mathematics?*). A home computer can factorize a 20-digit number, but attempting the same operation on a 100-digit number will generate severe struggles. As computers become more efficient, public key encryption can opt for larger numbers outside the factorizing powers of existing technology. But can we always do this?

In the McEliece cryptosystem, developed in 1978, the public key given out is not a single number or pair of numbers but rather a large *block* of numbers. For its use it is recommended that this block (termed a 'matrix') should have 644 rows and 1024 columns. This is used to encrypt the message, but the receiver has the added knowledge of how to split the block into three constituent blocks, corresponding to the factorization of the number into its prime factors. While the McEliece cryptosystem is unwieldy to use in practice it has a redeeming feature: it is not remotely crackable by existing methods.

Quantum codebreakers

Public key encryption might seem the end of the road as regards encryption techniques. But how much further can we go? Do we need to get even more secretive? Quantum computing is an emerging technology that *may* make the public key encryption prime number method redundant. While the 'classical' computer cannot find the prime factors of very large numbers, the quantum computer now on the drawing board offers an increase in efficiency that may unearth the primes. It would be a serious matter if it did, for it would mean that the encryption systems currently used for secure email and protecting web pages will be breakable.

So can we create an unbreakable code? The answer at present is 'yes', and we have: the McEliece cryptosystem currently remains secure, even against the theoretical quantum computer. But another answer has to be 'Watch this space.' In the battle between codemakers and codebreakers, the stakes are always rising.

IS MATHEMATICS BEAUTIFUL?

Music, art, golden numbers
and the Fibonacci sequence

*T*he idea of beauty in mathematics is perhaps a curious
one. We readily appreciate the beauty of a Leonardo
da Vinci drawing or a Mozart composition, but what is
beautiful about an erudite theorem, let alone pages of unknown
symbols? Mathematics can bring unexpectedness, challenge,
the satisfaction of understanding and the concise encapsulation
of a formula. But beauty? Two things are certain, though.
Mathematicians develop an acute sense of the beautiful, and the
history of the arts reveals a host of connections between ideas of
beauty and mathematical principles.

To say 'mathematics is beautiful' would mean little to a
schoolchild struggling with multiplication tables and percentages.
And when a mathematician goes into raptures about the beauty of
a mathematical construction or theorem, it often seems to require
a radical shift in perspective to see things in the same way. Can
the word 'beauty', with all its layers of often-contested meaning,
really be applied to mathematics?

The problems start with defining beauty. Many people
today would regard it as subjective. The philosopher David Hume
warned us that 'beauty in things exists merely in the mind which
contemplates them', a variant of Plato's 'beauty is in the eyes of the
beholder'. But aestheticians and thinkers have not always agreed.

For Immanuel Kant, beauty was an objective property and universal, and when it comes to catwalk models and leading men, there certainly seems to be a conventionalized category of 'the beautiful' in public life. One 19th-century mathematician, Arthur Cayley, who came up with octonions (see *Are Imaginary Numbers Truly Imaginary?*), looked on Plato as a guiding light but withdrew from this particular fray, noting: 'As for everything else, so for a mathematical theory: beauty can be perceived but not explained.'

However difficult to pin down, an aesthetic sensitivity to beauty has been a strong motivator for scientists and mathematicians. Paul Dirac, one of the greatest mathematical physicists of the 20th century, went so far as to write that 'it is more important to have beauty in one's equations, than to have them fit experiment'. It is an extreme view, perhaps, though Dirac was qualified to say it: he did, after all, come up with an equation that was one of the most significant events in the history of quantum mechanics.

We are not, however, entirely at sea in a world of imponderables. If we scratch the surface of the arts, from music to architecture and painting, principles emerge that are often considered vital to beauty and which are fundamentally mathematical – and perhaps none more so than symmetry and proportion.

Music and mathematics

Mathematics and music are often regarded as linked endeavours and abilities – indeed, skills in them have been attributed to the same areas of the brain. The basic physics of sound and the vibrating string was known to the Pythagorean scholars of Ancient Greece, who identified the link between musical scales and the ratios of whole numbers. Thus the value of the highest to the lowest pitch in the ratio $\frac{2}{1}$ makes up an octave of notes C, D, E, F, G, A, B, C, with the pitch in the higher C twice that of the C in the lower octave. In between, the pitch of the notes in the C-major chord – C, E, G – are in the ratios $\frac{5}{4}$ and $\frac{3}{2}$ in relation to C. Measuring in hertz, if the pitch in the lower C is 261 hertz, the pitch of G at a $\frac{3}{2}$ ratio is 391.5 hertz.

By adjusting ratios, music can generate harmony or discord, with the former a particularly central aspect of the pleasure we derive from it. Scholars of music have identified other mathematical principles in the way music is structured which govern its reception, and symmetry – a mathematical principle, if ever there was one – figures strongly. In music, strength and appeal often lies in the repeated pattern; a singular occurrence would sound impoverished by comparison.

In one curious case, mathematics and musical composition are clearly intertwined: in campanology, otherwise known as bell-ringing. Here, the practicalities of ringing bells in a certain order determine the nature of compositions. The mathematical patterns of the 'changes' have much in common with permutation group theory in modern algebra!

Beauty and shape

In discussing music, we have considered proportion and symmetry. Those factors are vital ingredients in the mathematics of shape and structure, too. Often they are most appealing at their simplest, as in an example of a humble rectangle and its properties.

'The pleasure we obtain from music comes from counting, but counting unconsciously. Music is nothing but unconscious arithmetic.'

GOTTFRIED WILHELM LEIBNIZ

What is the *most* beautifully shaped rectangle? We might say it is the square, a rectangle in which all sides are equal. This is arguable, for the square boasts symmetry and it has some special mathematical properties: the diagonals intersect at right angles, for example.

But there are other ways of analysing a rectangle, too, in which we appreciate its proportions. If we fold a rectangular piece of paper in half and cut along the fold we are left with two rectangular pieces.

If the longer and shorter side are in the ratio of $\sqrt{2} = 1.4142...$ to 1, the sides of the smaller rectangles will be in *exactly* the same proportion. This is the basis of the British 'A' method for sizing paper. Any piece of A-size paper is called a 'root 2 rectangle', because of its connection with the square root of 2.

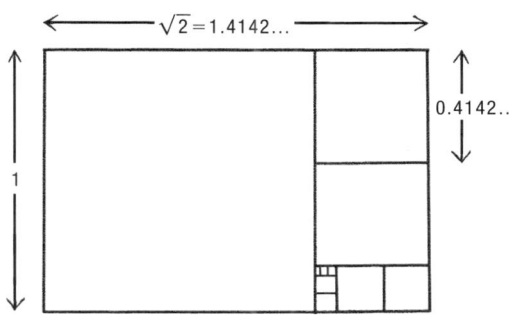

'TAKING SQUARES' FROM A ROOT 2 RECTANGLE

The root 2 rectangle can be analysed further by 'taking out squares'. If we take a square from the root 2 rectangle with actual sides 1 and $\sqrt{2} = 1.4142...$ we are left with a rectangle with short side 0.4142 and long side equal to 1.

From this we can subtract two squares of side 0.4142..., and then we are left with a rectangle of short side 0.1716... and long side 0.4142... From this, we can take out two more squares, and this process of subtracting two squares can be repeated *ad infinitum*. Another way of expressing this, mathematically, is that [1; 2, 2, 2, 2, 2, ...] is the *continued fraction* expansion of $\sqrt{2}$. There is, then, a satisfying symmetry to the root 2 rectangle; but it is not quite perfect.

Golden rectangles and golden ratios

Pride of place among *all* rectangles is the 'golden rectangle'. The ratio in the case of a golden rectangle is not $\sqrt{2} = 1.4142...$, but φ (phi), whose value is:

$$\varphi = \frac{1+\sqrt{5}}{2} = 1.618033...$$

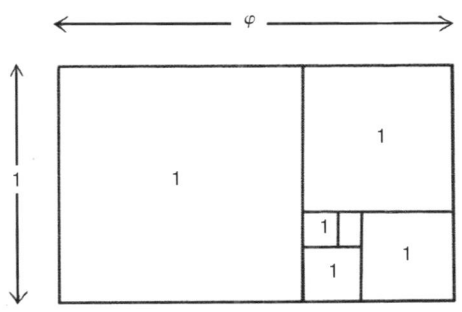

The number φ is called the golden number, or golden ratio, and it crops up throughout mathematics.

In a golden rectangle, if we subtract a square the remaining rectangle is *also* a golden rectangle. As we take out one square at each

A GOLDEN RECTANGLE AND ITS SQUARES

stage, the continued fraction expansion for φ is the rather special [1; 1, 1, 1, 1, 1...]

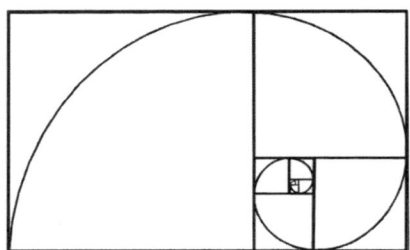

Drawing a curved line to connect the junction points of a golden rectangle's squares produces an elegant image. It approximates 'the logarithmic spiral', which highlights the rectangle's geometric properties, and it is a thing of beauty in itself.

For mathematicians, there is an inherent beauty in φ that needs no elaboration or application. It is enough unto itself. But moving beyond mathematics, the proportion that the golden number φ represents has been fundamental to Western culture. Because of its myriad properties, it has been called the 'divine proportion' after Luca Pacioli's *De divina proportione* (1509): he described φ in terms of the measurements of the human body, for instance the proportion of the body's height to the distance from the head to the finger tips.

In art, Leonardo da Vinci was much attached to φ as a proportion that occurs in nature. And, centuries later, the *Section d'Or* artists, breaking away from cubist orthodoxy, championed its mystical qualities and the sense of order that it has always symbolized. Several artists have been analysed from the golden-number perspective. The surrealist Salvador Dali's popular *Sacrament of the Last Supper* comprises a golden rectangle, and Dali makes use of mathematical figures in the painting. It is ironic, therefore, that the pre-eminent 'artist of the rectangle', Piet Mondrian, was one painter who somehow managed to escape the thrall of φ, his work representing no particular adherence to those proportions.

It is in architecture, however, where the golden ratio, sometimes referred to as the 'golden section' or 'golden mean', has had some of its most pervasive and enduring effects including the proportions of the Parthenon. In the 20th century, the

Swiss architect Le Corbusier explicitly celebrated mathematical proportions in his modernistic approach; his 'Modulor' theory proposed designs based on ground plans of golden rectangles.

Tessellations and geometry

An overt demonstration of mathematics in art occurs in regular decorative patterns of the kind called *tessellations*: arrangements of shapes or tiles, which go together in an exact fit on a flat surface. Symmetry plays a major part in the appeal of such designs.

Cultural factors within the Islamic tradition were influential in tessellations. Islamic tradition has generally militated against depictions of the Prophet Mohammed and other religious personages in art or in buildings. The result has been a vibrant and rich tradition of geometric patterning instead. In Spain, too, impressive tessellations are found in the 14th-century Alhambra in Granada, from the period of Moorish occupation. These particular examples are thought to have influenced the 'master' of tessellations in European art, Maurits C. Escher, when he visited Spain in the 1920s.

When a tiling, such as a tiling with hexagons, repeats itself, it is called 'periodic' tiling. Such tilings have *translational symmetry*: we can slide them over the flat surface and they will sit exactly over their previous image. With a non-periodic tiling, there is no translational symmetry. The pattern does not repeat itself, though there are tilings of this kind that still have 'rotational' symmetry (see *What Is Symmetry?*).

Several non-periodic tilings were discovered by Sir Roger Penrose, one of them based on just two shapes, the kite and the dart, where both are constructed from a rhombus – a shape with equal sides and pairs of parallel sides. And the sides of the kite and

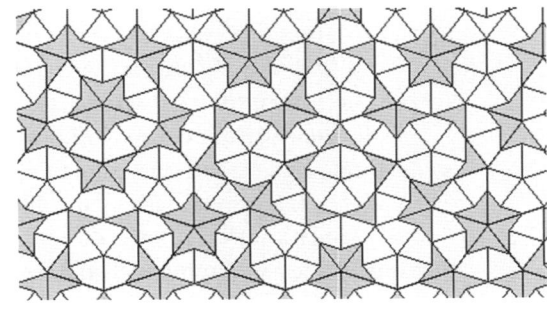

SOPHISTICATED EXAMPLE OF NON-PERIODIC PENROSE TILING

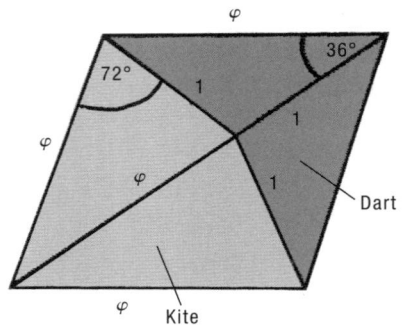

THE DART AND KITE, WITH 'GOLDEN' PROPORTIONS

dart have the added authority of being in the ratio of the golden number.

These non-periodic tilings, beautiful as they are, are also subjects of serious study in the new science of 'quasi-crystals'.

Beauty in numbers

So far we have looked at certain types of proportions, the part they play in the aesthetics of beauty and the objects that are produced as a result. But *within* mathematics, beauty is often perceived intrinsically: in the perplexing question, in the logical proof or an eloquent formula, and in a satisfying equation. Among these, the simplest are often the *most* appealing.

The famous (Pierre de) Fermat's last theorem is, in its simplicity, one of those beautiful questions. Formulated in the 1630s, it proposes that for the equation $x^n + y^n = z^n$ there can be no whole-number values for x, y and z for any number n greater than 2. There are whole numbers 6, 8, 9 that *almost* satisfy $x^3 + y^3 = z^3$, because $6^3 + 8^3 = 216 + 512 = 728$ and $9^3 = 729$, but we cannot find whole numbers that make the left-hand side of the equation *exactly* equal the right-hand side. The beauty of the theorem also lies in its pristine and puzzling quality, standing aloof from a proof that never seemed to come. It was more than 350 years later, in the 1990s, that the theorem was proved correct.

Then there are numbers themselves. To those of a mathematical mind, there resides a sheer beauty in numbers *per se*, in their varieties, arrangements and sequences. The 'Fibonacci sequence' is one of the simplest sequences of whole numbers, which has entered mathematical lore since Fibonacci (also known as Leonardo of Pisa) set it down in 1202. There is an academic journal devoted to its well-being, and new properties of the sequence continue to be discovered. It has even been claimed that composers of music have based compositions on the sequence.

The beauty of the standard Fibonacci sequence lies in its mathematics. To write it down, all we have to do is remember that it starts with the numbers 1, 1. The third term is the sum of these, so $1 + 1 = 2$, giving the sequence 1, 1, 2. The fourth term is the sum of the two *previous* terms, so $1 + 2 = 3$, thus giving 1, 1, 2, 3. Each term is just the sum of the previous two terms, and from this we can grow the Fibonacci sequence:

$$1, 1, 2, 3, 5, 8, 13, 21, 34, 55, 89, 144, 233, 377, 610, \ldots$$

There is more to it than that, though. The sequence is connected with the golden number φ. If we look at the *ratios* of each term divided by the previous one, we obtain the sequence

$$\frac{1}{1}, \frac{2}{1}, \frac{3}{2}, \frac{5}{3}, \frac{8}{5}, \frac{13}{8}, \frac{21}{13}, \frac{34}{21}, \ldots$$

which, in decimal form, becomes:

$$1, 2, 1.5, 1.666, 1.600, 1.625, 1.615, 1.619, \ldots$$

These decimal expansions are alternately *below* and *above* the value of the golden ratio $\varphi = 1.618033\ldots$ and as we take up more terms the sequence of ratios gets closer and closer to the exact value of φ. That kind of a correspondence is precisely the kind of ordered beauty that a mathematician can appreciate.

The beauty of invention

Great mathematicians have their own ideas of what, among their own discoveries, they find most beautiful, and it is not always the obvious or the most elaborate. Archimedes of Syracuse, who lived in the third century BC, is generally reckoned the finest mathematician of Ancient times. He contributed to both pure and applied mathematics, but the work he valued most highly was his discoveries about the sphere and cylinder, namely that when a cylinder covers a sphere, and is the same height and touches its sides, then the *surface area* and *volume* of the sphere are both $\frac{2}{3}$ that of the cylinder (see also *Is There a Formula for Everything?*). His discovery, without formulae, amazed him in its pure simplicity.

> '*Mathematics, rightly viewed, possesses not only truth, but supreme beauty — a beauty cold and austere, like that of sculpture, without appeal to any part of our weaker nature, without the gorgeous trappings of painting or music, yet sublimely pure, and capable of a stern perfection such as only the greatest art can show.*'
>
> BERTRAND RUSSELL,
> *The Study of Mathematics* (1902)

The sphere and the cylinder were seen as the building blocks of geometry in Ancient Greece. Over 2000 years later, the post-impressionist painters of continental Europe began to take this as their watchword, transforming representational art into geometric depictions. Paul Cézanne, a leading post-impressionist, advised one fellow artist to 'treat nature by the cylinder, the sphere, the cone', and the cubists such as Picasso and Braque took this approach to new heights.

In the 19th century, the eminent Carl Friedrich Gauss — accounted, with Archimedes and Newton, among the aristocracy of mathematicians — was able to unite several ideas in his geometric constructions, the resulting proofs of which are beloved by mathematicians for their elegance and power. He focused on constructions where the only equipment allowed was a straight-edge to draw straight lines (*not* for measuring distance) and a compass for drawing circles or parts of circles.

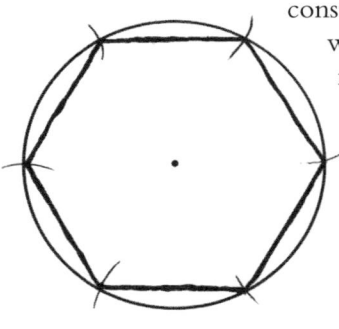

A HEXAGON DRAWN WITH COMPASS AND STRAIGHT-EDGE

It is not too difficult, in this fashion, to construct an equilateral triangle, a square, a regular pentagon with five equal sides, and a regular hexagon with six equal sides. To construct a hexagon, for instance, we choose a radius *r* and, with the compass points set a distance *r* apart, we first draw a circle. On the circumference of the circle we mark off arcs with the compass at the same setting, and after six such marks we arrive back where we started. We join up the marked points and we have our hexagon.

While Euclid had known how to make these shapes, he could not construct a regular figure with seven equal sides. Investigating this problem as teenager, Gauss proved it was not possible. Nor was it possible to construct a figure with 11 sides or one with 13 sides. It was not that Gauss failed to make these constructions, but that he *proved* they could not be made. Gauss proved a negative result.

Gauss's spectacular result had a positive aspect too. He constructed a regular figure with 17 sides. And he did more, for the number 17 has a particular form. It can be written in the form:

$$2^{2^n} + 1$$

That is because, with $n = 2$, $2^2 = 2 \times 2 = 4$ and $2^4 + 1 = 2 \times 2 \times 2 \times 2 + 1 = 17$. Gauss showed that, whenever this form is a prime number, the polygon with this number of sides can be constructed. The numbers of this form (called Fermat numbers) are prime numbers for the values of $n = 0, 1, 2, 3, 4$, being respectively 3, 5, 17, 257, 65,537, but no one knows if any prime numbers exist for other values of n. For $n = 5$, for example, the Fermat number is 4,294,967,297 but it is not prime, for:

$$4,294,967,297 = 641 \times 6,700,417$$

The question of beauty

There are, then, a myriad of reasons to associate mathematics with beauty. Certainly mathematics imbues music, art and architecture with proportion, symmetry and perspective. Are mathematicians so different to those who create in those fields? Mathematics is, in its own way, founded on aesthetic principles too. The 'eye of the beholder' notwithstanding, it would be a foolhardy sceptic who denied that there was beauty in mathematics.

CAN MATHEMATICS PREDICT THE FUTURE?
Mathematical models, simulations and game theory

*I*n 1865, as the American Civil War was drawing to a close, Abraham Lincoln was embarking on his second term as president of the fractured country. In his inaugural address he expressed his 'high hopes for the future', but wisely cautioned that 'no prediction in regard to it is ventured'. As it turned out, little more than a month later Lincoln's personal future was decided by an assassin's bullet, though his country went on to prosper. Would the application of mathematics have helped him evade his fate? The question sounds absurd, but mathematical means of 'prediction' have been applied to all manner of human behaviour and endeavour, in areas as diverse as politics, weather forecasting and share dealing.

We can easily pour scorn on the presumption that we can predict using mathematics, as if equations and symbols can become crystal balls. The global financial crisis that shook the world from 2008 appeared to deliver a withering verdict on mathematical hubris. For several years, financial institutions had been hiring mathematicians – they called them 'quants' – to forecast developments. In the eyes of critics, the results of all their esoteric mathematical models, using probability and statistics, was the subsequent financial meltdown, of a magnitude never witnessed before.

But we should be wary of throwing the baby out with the bathwater. It would be a mistake to take the excessive risk of the global banking system as suggestive of the *whole* field of mathematical modelling. In many, much quieter, ways, modelling has been going on for years, with far more plausible and accurate results. We might think twice before trusting a financial adviser again; but we are pretty confident that the solar eclipse predicted to the minute on 21 August 2017 will take place.

Modelling life

A mathematical model is a way of describing a real-life situation in mathematical language, turning it into the vocabulary of variables and equations. By making assumptions about what is important and ignoring some particulars that may be disregarded, the model aims to capture the *essentials* of a situation. A mathematical mechanism is set up that can then be validated, in order to check whether the model mirrors the real-life situation.

Mathematical models were traditionally seen as applying to physics. Newton's gravitational theory, which can model the motion of the planets, and Maxwell's equations (see *Is There a Formula For Everything?*), which model the interplay between magnetism and electricity, are principal examples. But today a wide range of mathematical techniques are called on to model situations in demography, economics, geography, psychology, biology, medicine, engineering – and beyond. In this way, for example, the growth of cities and populations has been put under the mathematical microscope, and so has the spread of disease in the population and even within the bodies of patients. The scope for modelling is ever widening.

> *'Prediction is very difficult, especially if it's about the future.'*
> NIELS BOHR,
> AND OTHERS

Given the very broad manifestations of modelling, it is unsurprising that a wide range of techniques has emerged, invariably making use of sophisticated mathematical notions such as probability, statistics, set theory and 'optimization' techniques.

In the light of relativity theory, Newton's gravitational theory may be seen as 'approximate', but the model derived from it remains sufficient to predict eclipses. In modelling financial markets, there is no clear-cut theory – perhaps indicative of the relative difficulty of predicting human behaviour as opposed to planetary behaviour. But here there is numerical data, and by using statistical techniques analysts can formulate equations, in an effort to evaluate underlying trends.

The whole field is now opened up by the increasing power of the computer. But behind the technology there remain the mathematical underpinnings that determine how *well* a model mirrors reality.

Modelling populations

The question of how *well* a model works is, of course, a vital though difficult question: by their very nature, the final answers to modelling questions might not be verifiable for some time, until real events catch up with them. The field of population studies is, however, a good example of an area where new techniques have built on earlier models to deliver far more accurate predictions.

In the 19th century, the Reverend Thomas Malthus formulated a famous mathematical model of population growth in his *Essay on the Principle of Population*. In it, he came to the pessimistic conclusion that an increasing population would outstrip the capacity of the earth to produce food, and he claimed that if epidemics, pestilence and plague did not carry off swathes of the population, then a 'gigantic famine' would achieve the same result.

In modelling terms, Malthus's theory assumed that a population doubles each 25 years, so, for example, population N would be $2 \times N$ after 25 years, and after a further 25 years would be $4 \times N$. At 25-year intervals the population *growth factors* are therefore:

$$1, 2, 4, 8, 16, 32, \ldots$$

Supposing that there is initially enough food (F) for the population, Malthus argued that it was inconceivable that food could increase as fast as the population, for there was a limitation on the land available for food production. Thus, he added F every 25 years, so that after the first 25 years there would be $F + F = 2 \times F$ available and after 50 years there would be $F + F + F = 3 \times F$, and so on. The food growth factors, according to this theory, are therefore:

$$1, 2, 3, 4, 5, 6, \ldots$$

The growth factors for population are in geometric progression (*multiplying* by 2 each time), whereas the growth factors for food are in arithmetic progression (*adding* 1 each time). Malthus therefore arrived at his conclusion that there would not be enough food for the burgeoning population.

Malthus's model was enormously influential in its time, but in hindsight the theory seems patently wrong. If it were used to predict, for example, the population of the United Kingdom *today*, we would find some strange results. From the population of 10 million around the year 1800, when Malthus was writing, his theory would have predicted a swelling tide of around 256 million by the year 2000 – over four times the actual figure.

To be fair, Malthus could not have taken into account such modern developments as birth control, immigration and emigration, and food technology. The Malthusian model is not a waste of time. It spurred more sophisticated approaches, in which better assumptions could be built in and the original model upgraded. We can now make the growth rate of a population *decline* as the population itself increases, and with this a ceiling is placed on the size of population (see also *Can a Butterfly's Wings Really Cause a Hurricane?*).

Forecasting and its limitations

In some areas of daily life, our keen appetite to know what lies ahead is challenged by the sheer fickleness of the phenomena investigated.

The movement of money and the state of the weather are good examples where modelling, strenuously applied, has its limitations.

In the financial world, stock-market prediction occupies much of a professional investor's time. Stock price indices in the major centres – the Dow Jones, the FTSE 100, the Nikkei – are all based on the amalgamation of a selection of stocks, and a combined figure is issued at the end of each day's trading. Predicting the index for the next day could mean the difference between gaining or losing millions. One difficulty, of course, is that financial markets are notoriously affected by the abstract notion of 'confidence' – predictions about human reactions to phenomena that can sometimes be quite unconnected to finance, but which may yet impact upon it.

The starting point, here, as invariably with mathematical modelling, is historical data. Just how the data is treated is where judgement comes in. A simple stock-price model could look at the previous days' closing figures. For example, if we know the indices are 3700, 3900 and 4100 on Monday, Tuesday, Wednesday respectively, these might be used as the basis of a prediction for Thursday. In the absence of other useful information, we could just average out the readings and suggest that our Thursday prediction would be:

$$\frac{3700 + 3900 + 4100}{3} = \frac{11,700}{3} = 3900$$

This technique is called the (simple) *moving* average method. This is because when the actual figure for Thursday is obtained, the data from Monday is dropped and the average of days Tuesday, Wednesday, Thursday provides the prediction for Friday. However, in a rising market an averaging method would always give a prediction that shows a dip, and it may be discarded by dealers who feel it is failing to reflect prevailing conditions. To partially compensate, greater weight can be placed on very recent history compared with history some days past, to try to reflect the trend better. The way to weight days is itself a matter of judgement, and different indices reflect different models. In real life, of course, dealers have specialist knowledge of a particular set of companies to feed into the overall prediction. Yet no forecasting model, of itself, can predict a stock-market crash.

Weather forecasting, too, remains an inexact science (see *Can a Butterfly's Wings Really Cause a Hurricane?*). Forecasts with any degree of accuracy are limited to just a few days ahead, and even then are liable to last-minute changes as real events unfold. The difficulty is that the Navier–Stokes equations that model the circulation of the atmosphere and the oceans, and hence dictate the weather, are impossible to solve exactly at the current time (see *Is There Anything Left to Solve?*). We have to rely on approximate solutions, but these are complex and involve a large number of computer calculations.

Outside weather predicting, even a simple everyday situation like how quickly a queue will form gives rise to equations that resist exact solutions in terms of a formula. Even if possible, a solution could apply only to an ideal situation on account of the assumptions that have to be made at the outset of modelling. Nevertheless, mathematicians have striven to tackle such circumstances by means of the method of *simulation*.

Simulation means experimenting with a mathematical model and making observations. How can we predict the number of supermarket checkouts that need to be operated at different times of the day? How should aircraft be stacked above an airport in different weather conditions? These are the types of questions approached by simulation. By varying parameters of the model – in the case of supermarkets, by times of the day; in the case of aircraft, by 'changing' weather conditions – the mathematical experimenter can 'see' what happens in various situations, even though they don't have an exact solution. It's not perfect, but it is a definite advance on guesswork and improvisation, which, in the case of hovering aircraft, could be catastrophic!

The lessons of games

Some mathematicians have adopted 'game theory' as a means of making predictions. This mathematical theory was pioneered by

John von Neumann in the 1940s, when he investigated what is called a two-person 'zero-sum' game. These games model the competitive situation between two players, who make rational choices based on a known table of pay-offs. The theory makes certain psychological assumptions that the participants act rationally to make decisions on what is best for them individually. The 'zero-sum' element means that what one player wins the other must, perforce, lose. This means that there is no point in the two players cooperating with each other.

There are also games where the zero-sum element is not assumed, and here the question of potential cooperation *is* brought into play. A well-known example of this sort is the 'prisoner's dilemma' problem designed by Albert Tucker. In this, two people, Alan (A) and Bruce (B), are picked up by the police on suspicion of highway robbery, but the police have insufficient evidence against them. They hold the prisoners in separate cells, so they cannot confer on how to plead. Sitting in their own cells, the two suspects act in their own best interests, and they cannot communicate with each other to coordinate a defence strategy.

The 'pay-offs' for Alan and Bruce are individual jail sentences, but these depend not only on how they respond to police questioning as individuals but on how they *jointly* respond – if one confesses and the other does not, the police will 'reward' the confessor by asking for a much lighter sentence for him than for his partner in crime: one year in jail as against ten years. But it would be very convenient for the police if both prisoners confessed, in which case a sentence of four years would be proposed for each and the case would be closed.

What should Alan do? If he confesses, the maximum penalty would be four years in jail, however Bruce chooses to plead. If he does not confess, the maximum jail term would be ten years. Alan, being a rational man chooses to confess, his strategy being the one that *minimizes* the *maximum* potential penalty, the so called 'mini-max strategy'. Bruce looks at the problem in the same way, chooses to confess too – and the

The prisoner's dilemma problem		Bruce	
		Confess	*Not confess*
Alan	*Confess*	4, 4	1, 10
	Not confess	10, 1	0, 0

ultimate result is that both end up with four-year jail terms. If Alan and Bruce had shared a cell and had been able to cooperate, the result would most likely have been different. In the absence of enough evidence against them, they would have both been able to plead 'not guilty' and walk away as free men.

Game theory and nuclear brinkmanship

The fictional fate of two petty criminals is one thing. In real life, the stakes were immeasurably higher during October 1962, when the Cuban Missile Crisis propelled the world towards the possibility of a Cold War nuclear conflagration. Yet this too has been analysed in terms of game theory, becoming a classic example of the basic principles for deciding actions. In 1962 it had been established by reconnaissance that the Soviet Union had set up launch sites in Cuba that would enable missiles to hit the US mainland. The level of threat was raised still further by the observation that Soviet transport ships were *en route* for Cuba, bearing missiles.

In game-theory terms, what were the options for US President John F. Kennedy and the Soviet leader, Nikita Khrushchev? It was a complex situation, each leader not knowing what the other would do. One way to analyse the situation, in retrospect, is to envision it 'on the back of an envelope' as a non zero-sum game. The United States had to choose whether it should blockade Cuba or mount an air attack and the Soviet Union whether to withdraw from Cuba or stay there. But what should be the pay-offs? How can we cast into numerical form the pessimistic outcomes? One way is to rank them numerically, with catastrophic as 1 and unsatisfactory as 5 with degrees of loss in between. So, for example, a pay-off of 5, 2 would mean unsatisfactory to the United States, but really bad for the Soviet Union, in this sense a 'US victory'. The numbers serve to *order*, and in some sense measure, how awful the situation is.

As happened in reality during those tense days, both countries stood back from the brink and managed to avoid a course of action that would have triggered war. The United States implemented a blockade, the Soviet Union agreed to withdraw the missiles from Cuba, and other concessions ensured a compromise was reached.

The Cuban Missile Crisis		Soviet Union	
		Withdraw	*Stay*
USA	*Blockade*	3, 3 (compromise)	2, 5 (US defeat)
	Air attack	5, 2 (US victory)	1, 1 (nuclear war)

Game theory has now been applied to cases involving more than two players and to games where cooperation and coalitions among players and groups are allowed, which, for example, can allow two players to form an alliance against a third. The study of game theory has advanced too, as seen in the influential work on the subject developed by US mathematician and economist John Nash, for which he was awarded a Nobel Prize in 1994.

Prediction: a health warning

It is not hard to think about situations where correct, timely predictions can avert disaster. Pandemics, tsunamis, earthquakes and volcanic eruptions fall into this category, but our abilities to tackle these threats vary considerably. Once these seismic events have begun, their spread can be plotted, but predicting time, place and magnitude *before* they have begun is the holy grail. Frequently, we must remain content with an uncertain prediction that hedges its bets in terms of probability (see also *Can Mathematics Guarantee*

Riches?), for example that 'there will be a thunderstorm with probability 70%', or 'there is about a 40% chance that the economy will grow between 1% and 3%'. The thorny issue of predicting climate change is a key area where predictions can only be made on the basis of probabilities, often controversial ones. Climate-change modellers, using mathematics and statistics as their basic tools, are rightfully cautious about their predictions. Only time will tell whether the facts marry up with any forecast.

While some mathematical models, such as Newton's, deliver near-perfect results given the nature of their contexts, others involve equations that give rise to unstable solutions. They are of the type that involve chaos theory (see *Can a Butterfly's Wings Really Cause a Hurricane?*). In these models, completely different solutions can be derived from initial values that are close together, and since those initial values are often uncertain in themselves, prediction becomes highly provisional.

It may sound as though we are saying that in the country of the blind the one-eyed man is king. To an extent, that is true, but his vision is improving all the time. The key to making a successful attempt at predicting the future is a mathematical model based on a satisfactory scientific theory. Mathematics and statistics, *by themselves*, are of little use, but in combination with scientific evidence they become stronger and more purposeful.

Can mathematics predict the future? Sometimes, and in the right contexts and with the right judgements. And there is one benefit we should not forget. Whether or not, ultimately, we can obtain usable predictions, constructing a mathematical model remains in itself a worthwhile exercise. When we identify variables and determine the ways they interact, the process puts us in the way of thinking about the problem more deeply and we are in a better position to analyse it. In particular we become more finely attuned to the validity of the assumptions we use, and without valid assumptions no prediction can expect to be accurate.

WHAT SHAPE IS THE UNIVERSE?

Topology, manifolds and the Poincaré conjecture

*R*amblers and hikers are accustomed to grappling with
topography, the distinguishing features of the landscapes
they walk in, such as hills and valleys. Mathematics has its
equivalent in 'topology', also derived from the Greek topos
('place'), and which is different from traditional geometry.
A precocious young mathematical subject, it makes use of set
theory and algebra, and it has achieved spectacular successes.
In tackling the shape of the universe, it faces one of the biggest
challenges of all.

In topology, mathematicians are not required to measure lengths,
angles, areas or volumes. Instead, they are more interested in
such questions as how individual points are connected, whether
surfaces have holes, or if shapes may be transformed into others.
Topology concentrates on the inherent properties of sets of
points, curves, shapes, surfaces and spaces.

Making the right connections

A glance at the diagrammatic map of the London or New York
Underground or the Paris Metro immediately tells us which
stations are connected to which other stations. A map it is, but
the positioning of the stations have only a very general bearing on
where they actually are, geographically speaking, and the length
of the lines joining them bear no relation to distance. The map

shows us how to get around the railway network, but not how far we travel or how long our journey will take. It tells us just what we need to know.

The railway map is an example of a topological *graph*. In topology, a graph (or network) is a diagram with 'vertices' (points), here the stations, which are joined to each other by 'edges', here the rail lines. In another terminological distinction, graphs in topology are not the same as the traditional graphs drawn on squared graph paper which illustrate how one quantity varies with another. In the context of vertices and edges, graphs constitute a completely different story.

The beginning of topology is often traced to the year 1735, when Leonhard Euler solved the 'Königsberg bridge problem'. Was it possible to walk around the easterly Prussian city of Königsberg (the modern Russian enclave of Kaliningrad), crossing each of its seven bridges exactly once? We might ponder it ourselves.

THE KÖNIGSBERG BRIDGE PROBLEM

However we plan the walk, we soon meet difficulties. We may be able to cross six of the bridges, but when we have done that we can't cross the seventh without doubling back over a bridge we have already crossed. There are two possible conclusions. Either we are not very good at planning routes, or it cannot be done. Euler came to the rescue and proved it was *impossible* to find a route. But he actually did much more than that. He produced a criterion that allowed us to check quickly whether or not a route is possible, for any city and for any configuration of bridges.

The revolutionary part of Euler's proof was his use of graph theory, for he replaced the actual map of Königsberg with a graph consisting of edges (bridges) and vertices (areas of land). One concept he used to advantage was the *degree* of a vertex, that is, the number of edges 'incident' on it, so for example the degree of b is 5 (see diagram). Euler's resulting theorem says that the bridges of a city may be traversed exactly once if, apart from at most two vertices, all vertices have an even degree.

From these 18th-century beginnings, graph theory truly got underway in the 19th century, and it has since found all kinds of applications. Today we operate the World Wide Web, where billions of connections (edges) between people (vertices) make up a graph that is changing every second.

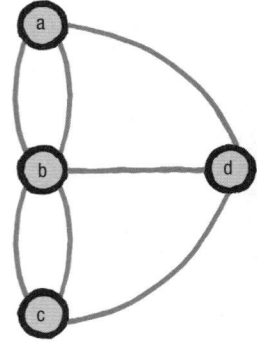

KÖNIGSBERG'S BRIDGES AS A TOPOLOGICAL GRAPH

Trees and their roots

In topology, a *tree* is a special sort of graph, where there is one path from any vertex to another (unlike the Underground system). A traditional family tree falls into this category, and it is a 'rooted' tree because it has a root vertex, the point of origin placed at the top of the tree. As well as its genealogical use, this type of rooted tree was the type chosen by Charles Darwin for his 'tree of life', showing the evolution of species. Rooted trees are useful for describing family trees where we are interested in grandparent, parent, child relationships as well as first cousins, second cousin once removed, and so on. We are familiar with a version of them in the directory/folder/file hierarchy adopted in computer systems for data storage.

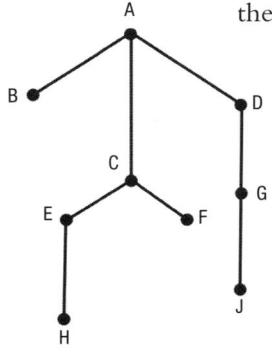

A TOPOLOGICAL ROOTED TREE

Some trees, however, are unrooted. They have no *preferred root*, and so for these there is no clear hierarchy. Counting the number of unrooted trees that can be created from a given number of vertices is a longstanding problem. We know that there are a total of six possible unrooted trees that can be created with six vertices. (It is possible to make *different-looking* unrooted trees by reorienting lines and points, but in terms of their connections there are only six structures.) But with more vertices, the variety of trees is staggering. For 21 vertices, for example, there are 2,144,505 different tree configurations.

The theory of unrooted trees has applications in organic chemistry, where they describe the make-up of molecules. The vertices describe the atoms, and the edges indicate the bonds between them. It is not just the number of atoms in a molecule that is important, but the manner in which they are arranged, their topological character. Arrange the number of atoms of the same type differently, and the result is a different chemical compound, a phenomenon known as structural isomerism.

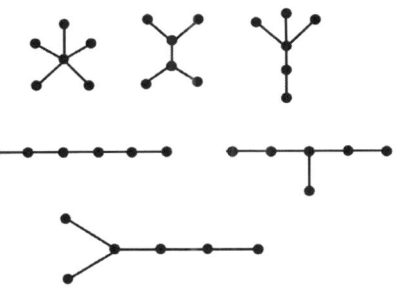

UNROOTED TREES WITH SIX VERTICES

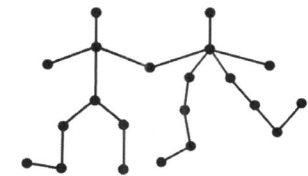

ONE OF **2,144,505** UNROOTED CONFIGURATIONS OF **21** VERTICES

In the 1870s, the mathematician Arthur Cayley calculated that there must be eight possible arrangements for the atoms of amyl alcohols with the chemical formula $C_5H_{11}OH$, even though only two of them were known to chemists at the time. It has since been found that all the theoretically possible amyl alcohols exist. Cayley had used the power of mathematics to theorize the existence of these chemical compounds long before they were found experimentally in the laboratory – just as his friend J.C. Adams had mathematically deduced the existence of the planet Neptune in advance of its astronomical identification. Both insights demonstrate the power that mathematics can sometimes bring to bear.

Knotty problems

Towards the end of the 19th century, topology expanded beyond vertices and edges as theorists tackled the problem of knots and how to classify them. A mathematical knot differs from a real knot in that there are no loose ends, so it resembles a single line interwoven in various ways. The mathematical interest lies in whether two knots are 'equivalent', that is, whether one can be manipulated into the other without cutting. For example there are two trefoil knots which are not equivalent to each other.

TREFOIL KNOTS THAT ARE NOT EQUIVALENT

Great tables of knots were assembled at the end of the 19th century, showing, for example, that there are 165 different knots with ten crossings. Quite recently, knot theory has been applied to the 'knottiness' of DNA. In practice, DNA exists in jumbled tangles, but replication is more efficient if the tangles are untied, and here mathematics makes its contribution.

Fixed points and hairy balls

Discs lie behind one of the most important results of topology. 'Brouwer's fixed point theorem' is named after its 20th-century Dutch originator, Luitzen E.J. Brouwer. We start with the observation that if a disc is placed on a table and rotated through an angle, then all the points of the disc move except one, the centre. This is not very surprising, but Brower's fixed point theorem says more and has wide applicability.

The theorem states that if we take the disc and trace its boundary on the table and then stretch it, shrink it, and fold it however we like, making sure that the final result lies inside the traced-out boundary, there will always be one point that sits above its original position, and is therefore classified as a 'fixed point'. There may be many of these points, but Brouwer said that there will be at least one. A party-trick statement of the theorem says that if a piece of flat paper is crumpled up and placed back on the table at the same place there will be still be a point in it that is lying over the identical point of the pre-crumpled sheet.

The theorem can be generalized to refer to a solid ball, and to cubes, and even to multi-dimensional objects. And it has spread widely. The American mathematician John Nash, investigating game theory (see *Can Mathematics Predict the Future?*), used it to prove the existence of equilibrium strategies when more than two players are involved (the so-called 'Nash equilibrium'). In economics, the fixed point theorem is used to show that certain prices have the effect of balancing supply and demand in a market economy.

Brouwer's theorem is an example of an *existence* theorem. It asserts the existence of something, but does not go beyond that.

Such theorems contribute to 'qualitative mathematical theory', in which calculations and measurements play little part. Another topological example is the notable 'hairy ball theorem'. It holds that if we have a ball with hairs sticking out from each point on it, then it is impossible to comb the hairs flat with a continuous movement without leaving at least one hair sticking up vertically (typically, at the centre of a swirl). We don't know *which* hair will stick up, only that one will. The theorem cannot apply to all geometrical objects, for in a doughnut shape it is perfectly possible to brush down all the hairs without leaving one vertical.

This apparently trivial theorem has implications for the study of wind patterns. We can think of the ball as the Earth and the combed down hairs as a measure of the wind blowing. The direction of a hair will be the direction of the wind at that point on the Earth's surface, and its length will measure the wind's force. The hairy ball theorem says there must be a place on Earth where there is no wind at all.

Manifold shapes and Möbius strips

A 'manifold' is a geometrical object whose *local* structure resembles an ordinary line segment in the case of a 1 dimensional object, or an ordinary disc in the case of 2 dimensions. This can be extended to a manifold of *n* dimensions. The sphere is an example of a 2 dimensional manifold, but a doughnut (torus) with one hole is another manifold and we could even have doughnuts with more holes than one. Manifolds are geometrical objects that may appear complicated when looked at in their entirety but when a portion of them is looked at they appear less so. It is the distinction between the global and the local picture, and in the case of the sphere this may be conceived as the difference between standing on a 'flat' piece of ground and seeing the entire spherical Earth from space. It is topology that handles the geometry of manifolds.

Furthermore, the sphere and the torus surfaces are *orientable* manifolds. If we attach an arrow pointing outwards (say) to a point on the sphere and manoeuvre the point around the surface, it will always point outwards, and the same applies to the torus manifolds.

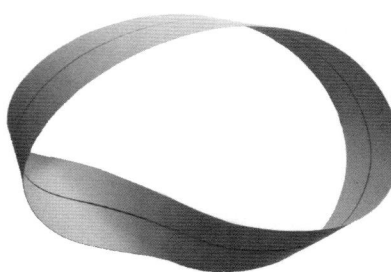

THE MÖBIUS STRIP — A MAVERICK MANIFOLD

There are, however, the maverick manifolds where this is not the case. The famous Möbius strip is one such, a two-dimensional surface formed from a strip of paper by giving it a twist and then gluing the ends together. It is one-sided, and if one attaches an arrow to it, pointing outwards at 90 degrees to its surface, and slides it along the strip, it will return to its starting point but pointing in the opposite direction.

The Klein bottle is another celebrated example of a manifold that is non-orientable. It can even be constructed from the Möbius strip! In the words of the limerick:

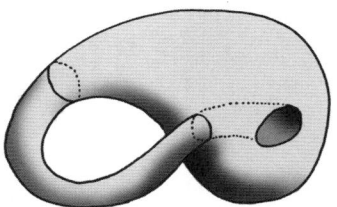

THE KLEIN BOTTLE

A mathematician named Klein
Thought the Möbius band was divine.
Said he: 'If you glue
The edges of two,
You'll get a weird bottle like mine.'

We would need four dimensions to see the Klein bottle properly, so the best effect is achieved by imagining it in three dimensions with intersections.

These manifolds – the sphere, doughnut, Möbius strip, Klein bottle with different characteristics – are all two-dimensional surfaces. The ordinary sphere, for example, is the boundary of the ball, and though the ball resides in three-dimensions the sphere is the surface of a ball, and so is two-dimensional.

If we rule out non-orientable manifolds, and such manifolds as infinitely long cylinders, and keep to smoothly shaped compact manifolds, it turns out that any such two-dimensional manifold can be deformed into either a sphere or a torus with r holes. This amounts to a classification of two-dimensional manifolds.

The story of the classification of three-dimensional manifolds is very different. Just about the simplest three-dimensional manifold is the three-dimensional sphere. By analogy with the ordinary sphere this is the surface of a four-dimensional ball, or equivalently the points in four dimensions which are equidistant from the origin. Given that we cannot easily visualize four dimensions we begin to appreciate the complexity of three-dimensional manifolds, when we think of the three-dimensional sphere.

So, three-dimensional manifolds have not been classified, although progress is being made. The rewards for the breakdown of the many types of three-dimensional manifolds would be great as links between them and seemingly disparate areas in both pure and applied mathematics are becoming apparent. And the great prize is the light that the classification would throw on the ultimate shape of the universe. The research programme to classify three-dimensional manifolds is of extreme importance.

Historically three-dimensional manifolds are intertwined with the famous Poincaré conjecture.

Simple connections and the Poincaré conjecture

Manifolds that are 'simply connected' are the subject of a celebrated conjecture put forward by Henri Poincaré. The terms of reference here can be demonstrated by imagining a point (P) on an ordinary sphere, attaching a loop of string to it, and moving the loop around the surface. However we do it, there is nothing to stop us pulling this loop down to the point while retaining contact with the surface. This works for any point and any loop, and it is this property that makes the sphere 'simply connected'. By contrast, a loop of string arranged around the hole on the torus does not have this property, because the hole would prevent the string being pulled down to the point where it is attached.

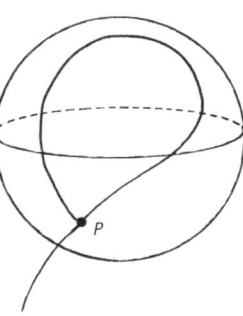

A TWO-DIMENSIONAL SPHERE
IS 'SIMPLY CONNECTED'

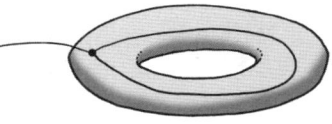

A TORUS (DOUGHNUT) IS NOT
SIMPLY CONNECTED

Since, as we mentioned, a regular two-dimensional manifold is equivalent to either a sphere or a torus with r holes, and the only one of these shapes that is simply connected is the sphere, we may conclude that any two-dimensional simply connected manifold must be equivalent to a two-dimensional sphere. So far, so secure. But what about the three-dimensional analogue?

This is the Poincaré conjecture: if a three-dimensional manifold is simply connected, must it be equivalent to a three-dimensional sphere? If we could classify the three-dimensional manifolds we would be in a position to answer this question, but the classification is not yet complete.

When Poincaré raised this question at the beginning of the 20th century, it became one of the great conjectures of mathematics. An n-dimensional variant of it was solved in the cases where n is greater than, or equal to, 5 in the 1960s, and after that came a proof in the case $n = 4$. The one unsolved case was the three-dimensional one, $n = 3$. Sensationally, it was recently solved in the affirmative by the reclusive Russian mathematician Grigori Perelman. The techniques used were quite outside the usual area of topology, depending on the mathematics of heat diffusion. Perelman was judged to have won the Clay Institute's million-dollar prize on offer for solving the Poincaré conjecture, but so far he has shown no interest in accepting it. He appears to have all he wants, values his privacy and is uninterested in fame.

The shape of the universe

Topology rose to prominence at the beginning of the 20th century and, richly expanding, has shown itself to be a useful branch of mathematics. But what about the big question? If topology is the geometry of the inherent properties of shapes, surfaces and spaces, what is the shape and size of the universe?

On a human scale, the physical universe is certainly vast. In our tiny part of it, made up of our solar system, the Sun is the nearest star to us, at a distance of 8 light minutes away, the time it takes light to travel from the Sun to us. The next nearest star, the red dwarf Proxima Centauri, is about 4.24 light years away. Most of the stars we observe are in the Milky Way, this comprising our own galaxy, but beyond this there are literally billions of galaxies of similar size. Edwin Hubble showed that these galaxies are moving away from us, and the further away they are, the faster they are moving. This is in agreement with predictions made from Einstein's equations of general relativity, as modified by Aleksandr Friedmann. It is now generally accepted that the universe was created at the moment of the Big Bang and is still expanding. But what is its shape?

The question of the shape of the universe returns us to the idea of the manifold, that is, the local perspective of the observable universe compared with the larger perspective of the entire universe. The local geometry involves curvature, for according to Einstein's theory of general relativity the presence of large masses bends space–time. We need to invoke the geometry theorized by Bernhard Riemann (see *Where Do Parallel Lines Meet?*), in which curvature may vary from place to place. That leaves the large question of the shape of the whole universe, and here we need topology to help us.

The physicist Ed Witten attested to the importance of topology in considering this question, noting that 'topology is the property of something that doesn't change when you bend it or stretch it as long as you don't break anything'; in Einstein's general relativity the structure of space–time can change but not its topology. The structure of the four-dimensional space–time for the universe as a whole involves determining the three-dimensional manifold that describes it.

Another way of saying this is that the theory of topology will be an essential element if we are ever to comprehend the shape of the universe.

WHAT IS SYMMETRY?

Patterns, dualities and the fundamental nature of reality

S ymmetry is all around us. Buildings have it, flowers have it, the human torso has it. We easily recognize symmetry, for the brain has a predilection for appreciating regularity in things observed. There has always been, since ancient times, a general agreement that symmetry in nature equates to beauty, as exemplified by the startlingly various patterns of snowflakes. But how can we define symmetry? 'Balance' and 'pattern' often come to mind, and symmetry has even been described as 'supreme equipoise'. But while those descriptions appeal to our aesthetic and artistic sensibilities, we need a different kind of language for mathematics.

It is difficult to arrive at a single definition of symmetry. Undoubtedly there are properties that most people would associate with the term: the idea of pairings, a sense of a shape being replicated, a recurring pattern. In mathematics and the sciences, symmetry comes in various guises – we are talking about a diverse notion.

Rotational and reflective symmetry

The symmetry of shapes is often discussed in terms of *rotational* and *reflective* symmetry. In rotational symmetry, we are interested in locating an axis around which a rotation would, to all appearances, leave the object looking the same (or, as mathematicians would say, 'bring the object into coincidence with itself'). A disc in two-dimensions has this symmetry, as does a sphere in three dimensions. Both a disc and a sphere possess an infinite number of rotational symmetries.

More interesting are the shapes that have a limited number of symmetries, such as, to take an example, a two-dimensional square with a windmill pattern. It has rotational symmetry, because if we rotate it by 90° in an *anti-clockwise* direction (say) about an axis passing through its centre, it is 'brought back into coincidence with itself' and the shaded areas of the windmill pattern line up again. The square yields four rotational symmetries, turning through 90°, 180°, 270° and finally 360° (which last is equivalent to 0° or no rotation at all).

The windmill square does not, however, possess *reflective symmetry* about any of its axes. However, if we removed the shaded windmill areas leaving the square plain, it would have reflective symmetry. In that case, from a mathematician's perspective, there are four lines of reflective symmetry: a horizontal line, a vertical line, a 'dexter' diagonal line (top left corner to bottom right corner) and a 'sinister' diagonal line (top right corner to bottom left corner). Including both the reflective and rotational symmetries, there are a total of eight symmetries for the plain square.

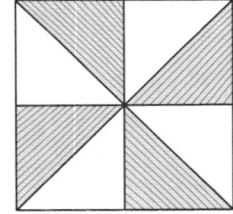

A SQUARE WITH WINDMILL PATTERN

The symmetry of invariants

If we measure the area of our square, it will be the same after we have rotated it or reflected it – unlike the case if the square had been stretched. The area is an example of an *invariant* if we restrict ourselves to certain symmetries.

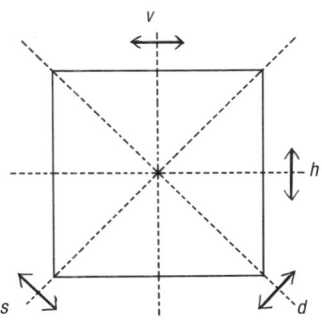

THE EIGHT SYMMETRIES OF A PLAIN SQUARE

This leads to the way physicists view symmetry, in terms of the quantities or qualities of physical systems which are unaltered by appropriate symmetries (such as changing a frame of reference).

> *'Symmetry, as wide or as narrow as you may define its meaning, is one idea by which man through the ages has tried to comprehend and create order, beauty, and perfection.'*
>
> HERMANN WEYL

So physicists identify symmetry with invariance. Galileo observed that an object falls vertically whether it is dropped in the laboratory or on a ship moving at a constant speed, and in this there is symmetry. The equations of Newton's laws of motion obey the same symmetry. Einstein went one stage further, when he placed this sense of symmetry squarely at the centre of his theory of special relativity, by declaring in advance that the speed of light was an invariant and was the same for *all* frames of reference.

A physicist would say that equations of motion are invariant when the variables are transformed, the transformation corresponding to different frames of reference. In this way the laws of physics in the Galilean, Newtonian and Einstein schemes are symmetric.

Measuring symmetry

Mathematicians and physicists approach the 'measuring' and manipulating of symmetry by way of the discipline called *group theory*. It originated in the early 19th century with the young mathematician Évariste Galois's work on the theory of equations. Before his early death at the age of 20, in a duel, he sketched out ideas for one of the most beautiful theories in all of mathematics (and which has been influential in physics and chemistry too). What then is a mathematical group?

With the windmill patterned square we have an example. Returning to this square, we denote a rotation through 90° by r, and apply it once again, we have effected a rotation through 180°. This combining of two rotations we denote by $r \times r$, which for short we can write as r^2. Rotating by a further 90° is the operation r^3, while r^4 will be a rotation of the square through 360° and back to where we started.

The combination of rotations can be conceived as 'multiplication', and this makes it possible to construct a 'multiplication table' of the symmetry of rotations. For instance, if we wish to rotate the square by 270° and then by 180° this would be denoted as $r^3 \times r^2$. This rotation through 270° + 180° = 450° in effect means a rotation through 450° − 360° = 90°, in which case we could write it as $r^3 \times r^2 = r$. The special symmetry that rotates the square through 0° is denoted by e, and so we can write $r^4 = e$.

×	e	r	r^2	r^3
e	e	r	r^2	r^3
r	r	r^2	r^3	e
r^2	r^2	r^3	e	r
r^3	r^3	e	r	r^2

THE MULTIPLICATION GROUP TABLE FOR ROTATIONAL SYMMETRY OF A SQUARE

The *reflective symmetries* of a square − horizontal, vertical, dexter diagonal and sinister diagonal − are labelled h, v, d and s, and they can be combined with the rotational symmetries. A horizontal reflection followed by a vertical reflection, that is $h \times v$, is actually equivalent to a rotation of the square through 180°, and so $h \times v = r^2$.

A complete multiplication table for the enlarged 'group of the square', consisting of the eight reflective and rotational symmetries, can be created.

Groups and symmetry

In general, a group consists of 'elements', of which any two can be combined together to give an element of same type. In the case of symmetries, two symmetries combine together to give another symmetry. Certain rules apply, which a combination must satisfy: for each element there should be another element in the group that reverses it; and when three elements are combined together it does not matter which two we combine

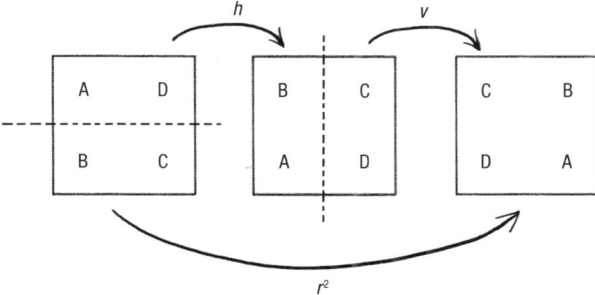

REFLECTION AND ROTATION EQUIVALENCE OF $h \times v = r^2$.

first. Under these modest requirements, an extensive theory has been constructed by legions of mathematicians.

When a group has a finite number of elements, the number of them is the *order* of the group, so our plain square yielded an order of eight symmetries. The multiplication table for this group shows the sub-groups, $\{e, r, r^2, r^3\}$ for rotational symmetry, $\{e, r^2, h, v\}$ for the reflections in the horizontal and vertical and $\{e, r^2, d, s\}$ for the reflections in the diagonals. The multiplication tables for these last two groups yield the same results, that is, despite the different letter designations, they have the same multiplication table as each other, and this type of group is called the 'Klein four-group'. A Klein four-group and the rotation group $\{e, r, r^2, r^3\}$ are the only types of group possible in a group with four elements.

A problem in the theory of groups is finding the number of different groups there are for a given order. There are two groups with order 4, for example. If the order of a group is a prime number there is only one group. If the order is a composite

$\times$	e	r	r^2	r^3	h	v	d	s
e	e	r	r^2	r^3	h	v	d	s
r	r	r^2	r^3	e	d	s	h	v
r^2	r^2	r^3	e	r	v	h	s	d
r^3	r^3	e	r	r^2	s	d	v	h
h	h	s	v	d	e	r^2	r^3	r
v	v	d	h	s	r^2	e	r	r^3
d	d	v	s	h	r	r^3	e	r^2
s	s	h	d	v	r^3	r	r^2	e

DIHEDRAL GROUP OF SQUARE

(non-prime) number, the problem is more interesting. For example there are five different groups of order 8. One is the group of the square, as we have seen; another is formed from quaternion 'imaginaries' (see *Are Imaginary Numbers Truly Imaginary?*), which are 1, i, j, k, -1, $-i$, $-j$, $-k$.

The symmetry of a shape is measured by its groups of symmetries, so our plain square is 'measured' by its full group of order 8 (the dihedral group) and by its Klein four-group and the rotation group.

Mirror symmetry

We can expand the repertoire of symmetries by including 'translation' and mirror symmetries. When we considered reflective symmetry we were thinking of horizontal, vertical and diagonal mirrors but now we are going to stand in front of a mirror. In mirror symmetry, as one would expect, one confronts an object with its mirror image. Of course, mirror images often involve reversals, depending on the object in question. A plain sphere held up to a mirror would produce no obvious variation in its mirror image; it has mirror symmetry. But a right hand held up in front of a mirror produces an image of what appears to be a left hand. Where objects in 'mirrorland' are different from their originals in this way they are known as *chiral*. In chemistry, the chirality property is important. In one of its forms, a molecule of a hydrocarbon in the skin of a lemon naturally smells of lemon, but the 'chiral twin' molecule, named limonene, smells of oranges. In physics researchers were astounded by the presence of chirality in the study of the so-called 'weak interactions' of particles. They eventually came to the conclusion that, in working out the theory of weak interactions, mirror reflections should be barred.

The symmetry of dualities

Mathematicians are fond of the concept of dualities and twins, which defines a kind of symmetry that can be applied to the symbols, theorems and algebra of their trade. Its force lies in the language and terms used to describe mathematical situations.

The geometry of the triangle provides a straightforward example of duality. There are three points (say *A*, *B* and *C*) and three lines (say *a*, *b* and *c*), and there is a symmetry between points and lines, so that we can say 'any two points determine a single line' as well as, transposing the terms, 'any two lines determine a single point'. There is a duality in the statements. The usefulness of this geometric duality is that a theorem written in terms of points and lines can be *automatically* turned into a theorem written in terms of lines and points; they occur in pairs.

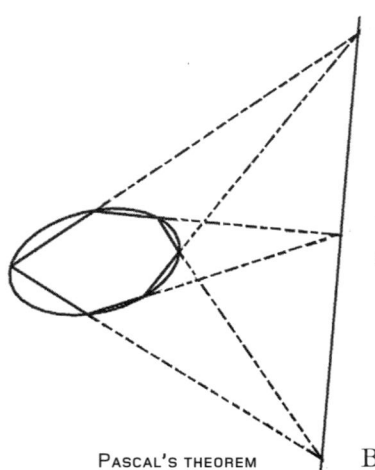

THE FANO PLANE

A richer example would be the seven-point and seven-line geometry known as the Fano plane (see *Are Imaginary Numbers Truly Imaginary?*). The seven lines include the circle as one of their number. In this geometry, the theorem that 'any three points determine a unique line' has the dual theorem 'any three lines determine a unique point'. As with the ordinary triangle, the geometry consists of a definite number of points and lines, but duality also occurs in regular geometry.

The most famous *twin* theorems in traditional geometry are Pascal's theorem and Brianchon's theorem. They were individually discovered more than 150 years apart, so the recognition that they were dual to each other came in retrospect. Pascal's theorem is concerned with a hexagon bounded by an ellipse, and it states that:

> Three *points* of intersection of the lines formed from opposite sides of the hexagon lie along a straight *line*.

PASCAL'S THEOREM

Brianchon's theorem states that:

> Three *lines* joining the points of intersection of tangents on opposite sides of the hexagon meet at a *point*.

In geometry, the symmetry of duality is quite common, and it is present in the five so-called 'Platonic solids' of the Ancient Greeks – the tetrahedron, cube, octahedron, dodecahedron and icosahedron. For example, if we take a cube and identify the midpoint of each of its six faces, and then join these points

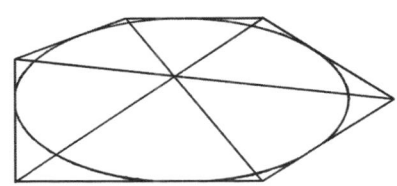

BRIANCHON'S THEOREM

to those on *adjacent* faces, the lines create a solid octahedron inside the cube. In other words, the octahedron is dual to the cube, and the duality is achieved by interchanging faces with vertices. A cube has six faces and eight vertices; an octahedron has six vertices and eight faces. In a similar way, the 12-sided dodecahedron is dual to the 20-sided icosahedron, which only leaves the tetrahedron, which is the dual of itself.

Duality is everywhere in logic and algebra, and it brings out the symmetry of language, and the way simple connectives like OR and AND are used. In logic, for instance, OR can be used to combine logical propositions such as 'I have a dog', 'I have a cat' into 'I have a dog OR I have a cat.'

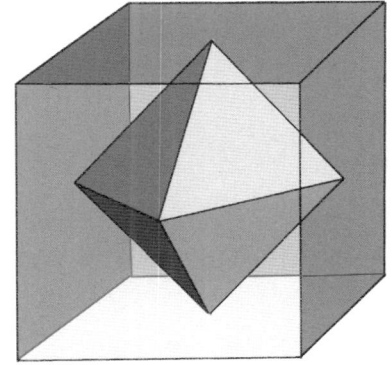

THE DUALITY OF CUBE AND OCTAHEDRON

(In everyday language 'or' usually rules out both occurrences but mathematics uses the inclusive 'OR', so the combined proposition could mean I have both animals.)

In logic basic propositions are built up into more involved ones with words like OR, AND and NOT, so we might have

'I do NOT have a dog OR a cat'

which is the same as saying:

'I do NOT have a dog AND I do NOT have a cat'

179

Solid	Duality	Faces	Vertices	Edges
tetrahedron	self-dual	4	4	6
cube	dual of octahedron	6	8	12
octahedron	dual of cube	8	6	12
dodecahedron	dual of icosahedron	12	20	30
icosahedron	dual of dodecahedron	20	12	30

THE SYMMETRICAL PROPERTIES OF THE FIVE PLATONIC SOLIDS

In logic the symbols $\vee$, $\wedge$, $\neg$, are used for OR, AND, NOT so the first proposition can be written in the form $\neg(P \vee Q)$ and the second $\neg P \wedge \neg Q$ (where P, Q stand for the individual propositions). Their equality gives De Morgan's law:

$$\neg(P \vee Q) = \neg P \wedge \neg Q$$

The duality symmetry here is between OR and AND, so in any true statement we can swap them. In the case of De Morgan's law we have another one, a second De Morgan's law:

$$\neg(P \wedge Q) = \neg P \vee \neg Q$$

A similar set-up occurs in set theory (see *How Big Is Infinity?*). A set is a collection of objects, and the *union* of two sets A, B is the collection of objects which are in one set or the other, or in both. The *intersection* of two sets A, B is the collection of objects which are in both A and B. The *complement* of a set A is the set of objects not in A.

Denoting union, intersection and complement by $\cup$, $\cap$, ^{c}, there is duality between $\cup$, $\cap$, and we get a corresponding pair of De Morgans laws for set theory

$$(A \cup B)^c = A^c \cap B^c$$

and its 'twin':

$$(A \cap B)^c = A^c \cup B^c.$$

Dualities frequently occur in pairings. And if we take the dual of the dual – for example, if we find the dual of the second De Morgan law – we return to the first De Morgan law, an added feature of duality symmetry.

Sub-atomic symmetry

Cutting-edge physicists are attempting to create a unified field theory that would set down laws to link the basic four interactive forces of nature and they clearly recognize the importance of symmetry in their endeavour.

Physicists have their own idea of 'pairing', too, which manifests itself in a highly speculative theory in particle physics that advances the idea of *supersymmetry* in the context of the fundamental particles known as 'bosons' and 'fermions'. Supersymmetry says that to each boson there is a corresponding fermion, and vice versa. The Higgs boson is the most famous of all the different types of bosons, but it has only been hypothesized by the standard model of particle physics: no one knows whether it exists or not. There is hope that the Large Hadron Collider, built underground in Switzerland, will settle the matter. If this so-called 'God particle' does appear, then the symmetries used to explain particle interactions in the standard model will be validated and we will be much much closer to understanding the fundamental nature of reality.

The diversity of symmetry

Mathematicians constantly deal with symmetry in all its forms, and use it as a guide to the correctness of their theories. For them, group theory often holds the key, and through groups symmetry can be defined and exploited to give a wider view of the mathematical landscape. But what, ultimately, is symmetry? Perhaps the best answer has to be a collective one. It is a manifestation of similarity, variously defined, from the rotations and reflections of squares, through the dualities of algebra, to the twin theorems of Pascal and Brianchon – and beyond, into (if the physicists are right) the very nature of matter.

IS MATHEMATICS TRUE?

From Plato's reality to Gödel's incompleteness theorems

The 'quintessence of truth' is how the poet Coleridge described mathematics in a letter to his brother in 1791. He was reflecting a view held by the Ancient Greeks and handed down over the centuries, which Coleridge had absorbed during his studies at Cambridge. But he was on the cusp of a revolution, not only in his own sphere of literature, but also in mathematics. Its status as 'beacon of truth' was about to undergo intense scrutiny and challenge.

It is tempting to think that numbers are just numbers: straightforward, uncontroversial – banal even – and inviolable. Are we not agreed that $2 + 3 = 5$ is true and cannot equal any other number? Are there not 180° in a triangle and not 181° or any other value? These mathematical 'facts' are ingrained in our mental equipment from a very early age and we don't usually question their truth. We automatically accept that other areas of thought are riven with debate and contention, but mathematics seems somehow above that. That kind of purity is what Coleridge had in mind, and until the beginning of the 19th century that reflex was indeed the orthodox position, originally imbued into hundreds of university-trained young men as they absorbed the learning of Ancient Greece.

Today, some mathematicians solve highly technical problems or construct erudite theories high on the edifice of mathematics, and they might not question the foundation of mathematical knowledge that supports them. But there are other mathematicians, with a philosophical bent, who do consider the question 'What makes a mathematical fact true?' They work in the basement of mathematics, and they are concerned with shoring up the foundations, making mathematical propositions credible.

What actually is the nature of mathematical truth? Are mathematicians like scientists discovering pre-existing phenomena? Or are they, in fact, more like artists starting off with a blank canvas which they can fill as they choose?

When we first learn mathematics, there appears to be no leeway. Subtraction, multiplication, addition and division produce, each time, one correct answer. This would suggest that mathematics is an unambiguous language for describing the world. Accordingly, the job for the mathematician is to locate the mathematical truths, to find what is 'out there', using solid principles and impeccable logic, and enter them into the collective book of mathematical knowledge. As truths accumulate, our knowledge grows, and the book of mathematical lore swells. This air of definitiveness applies equally to geometrical facts, the authority for over 2000 years being Euclid's *Elements*, a book that begins with self-evident truths and theorems and follows this with deductions from them.

The truth-seekers

It is a reasonable guess that most mathematicians see themselves as discoverers of pre-existing truths. Mathematical truth-seekers follow in the footsteps of Plato, who talked about ideal forms that exist in a separate realm, a 'Platonic world'. Numbers on their own may be regarded as Platonic forms – unchanging, eternal and timeless.

The distinction between what we discover on earth and what is true of the ideal forms is encapsulated in Plato's famous allegory of the cave, in which prisoners sit in a cave with their backs to a fire and can only see shadows of the realities that exist outside. The shadows inside the cave constitute the only kind of reality the prisoners can see until, by gaining knowledge, they are brought to the opening of the cave, whereupon they can perceive the 'really real'.

In our own flawed world we can draw all the triangles we want, but each one is only an approximation to a Platonic triangle form composed of points with no size and lines with no thickness. This is what Euclid tells us, and understandably so, for he learned mathematics in Athens from the pupils of Plato. In the *Elements*, Euclid demonstrated theorems about triangles with two equal sides (isosceles triangles), proving that they have equal base angles; but in so doing he was proving a fact not about the sketches we might see in a book, but about the 'real' triangle in Plato's realm. The real things can only be seen, said Plato, 'with the eye of the mind'.

The Platonic philosophy has exercised a powerful influence on mathematical thought down the ages. It encourages a spiritual dimension to the mathematical quest. The English mathematician G.H. Hardy subscribed to this view, and in his *A Mathematician's Apology* he wrote: 'I believe that mathematical reality lies outside us, that our function is to discover or observe it, and that the theorems which we prove, and which we describe grandiloquently as our "creations", are simply our notes of our observations.'

> '*Student:* *And all arithmetic and calculation have to do with number?*
> *Teacher:* *Yes.*
> *Student:* *And they appear to lead the mind towards truth?*
> *Teacher:* *Yes, in a very remarkable manner.*
> *Student:* *Then this is the knowledge of the kind for which we are seeking.*'
>
> A PLATONIC DIALOGUE

The truth creators

The possibility of mathematicians being creators or inventors of mathematical knowledge, rather than merely finders of it, was raised towards the end of the 19th century, as mathematicians re-envisioned what they were doing. Do numbers exist in nature? In one sense, clearly not, for they are symbols and aspects of language. Proceeding from that, it could be said that we create the mathematics we need. Such thinking led to a philosophy of mathematics different from that based on Plato.

The German Leopold Kronecker famously said: 'The natural [i.e. whole] numbers come from God, everything else is the work of man.' Kronecker wished to reduce everything to arithmetic and ultimately back to the whole numbers. He believed that all mathematical reasoning had to take place in a finite number of rigorous steps, and he only allowed mathematical objects to exist if they could be constructed from the whole numbers this way. For this reason, the constant π did not exist because, for Kronecker, we could not know *all* the terms in its infinite expression:

$$\pi = 1 - \frac{1}{3} + \frac{1}{5} - \frac{1}{7} + \frac{1}{9} - \cdots$$

When Ferdinand von Lindemann proved that π was a transcendental number (see *Which Are the Strangest Numbers?*), Kronecker complimented him on a beautiful proof, but added that it had proved nothing since π did not exist. This was an attack on the traditional view of mathematics, as handed down by the Greeks, where π was established as the number associated with the circle and this was enough evidence for its existence.

The 19th century witnessed a rapid growth in the academic study of mathematics, expressed in the creation of university chairs and a general professionalization of the subject, accompanied by an efflorescence of new journals. The combined effect was to turn a trickle of mathematical results into a flood. New developments continued to throw doubt on the view of mathematics as the unearthing of 'truth', and there was a reappraisal of its foundations.

> *'The fact that all Mathematics is Symbolic Logic is one of the greatest discoveries of our age; and when this fact has been established, the remainder of the principles of mathematics consists in the analysis of Symbolic Logic itself.'*
>
> BERTRAND RUSSELL,
> *Principles of Mathematics* (1903)

Mathematicians wrote about strange new multiplication where $a \times b$ and $b \times a$ were *different* (see *Are Imaginary Numbers Truly Imaginary?*), algebra took off in new directions and new geometries different from Euclid were proposed (see *Where Do Parallel Lines Meet?*).

Increasingly, mathematicians were becoming self-conscious about their subject and its definition. The American mathematician Benjamin Peirce offered the view that 'Mathematics is the science which draws necessary conclusions.' Peirce aligned mathematics with logic, and he was not the only one. In his *Principles of Mathematics* (1903), Bertrand Russell famously described pure mathematics as 'the class of all propositions of the form "p implies q"'. For Russell, it was the logical structure that was important and not the actual truth of p and q, a consideration that prompted his later summary of mathematics as 'the subject in which we never know what we are talking about, nor whether what we are saying is true'.

A new direction

Where did all this leave mathematics? In some places it was a struggle between neo-Platonists and Young Turks, with the once unambiguous 'truth' dissolving into a bewildering array of man-made, even competing, truths. In Germany, Georg Cantor created a theory of sets, and there was hope that it would give mathematics a proper foundation.

A set is a *collection* of objects. In mathematics, if we want to know something about the whole numbers, for example, we are referring to the *set* of whole numbers. It appears to be a trouble-free concept, but Russell almost destroyed the theory at birth with a single blow by producing an internal contradiction, a result known as Russell's Paradox.

To investigate Russell's Paradox we need a set which has the capacity for being a member of itself. We could begin by taking a set, say the set A, of all abstract things. The set A, being a collection of abstract things, is also an abstract thing *itself*, and so A is a member of itself. That relationship is written as $A \in A$, the sign '$\in$' denoting 'membership'. But if we now consider the set S, which is defined as the set of all those sets that are *not* members of themselves, we obtain the paradox. It works in this way. Let us suppose that S is a member of S (that is, $S \in S$): it must logically satisfy the defining relation of set S – but that is that S is *not* a member of itself. Mathematicians write that as $S \notin S$. However, if we start out with $S \notin S$ then, by this property, S does meet the definition of S so we are returned to $S \in S$.

The conclusion is Russell's Paradox: that S is a member of itself if, and only if, S is *not* a member of itself! It resembles the well-known barber paradox, where the village barber is instructed to shave all the men in the village who do not shave themselves. Should he shave himself? If he does, he should not; while if he does not, he should. This kind of internal contradiction in any theory needs to be avoided.

One way out of this difficulty was to frame a list of axioms for set theory that would outlaw the kinds of sets involved with paradoxes. One list was derived by Ernst Zermelo and Abraham Fraenkel and supplemented by an 'axiom of choice' during the first decades of the 20th century. This was generally accepted as a useful list, and it provided a bulwark against the paradoxes. But questions remained. Was the system defined by this list of axioms free of all contradictions? Were there other unforeseen paradoxes waiting to spring out?

> *'... there is [a] considerable amount of controversy in mathematics. Pure mathematicians disown the proofs of applied mathematicians, while logicians in turn disavow those of pure mathematicians. Logicists disdain the proofs of formalists and some intuitionists dismiss with contempt the proofs of logicists and formalists.'*
>
> IMRE LAKATOS, MATHEMATICS,
> *Science and Epistemology*

The formalists

In an attempt to provide a solid foundation for mathematics, the German mathematician David Hilbert proposed that 'classical' mathematics should be treated in an axiomatic form, and from the axioms theorems could be logically deduced by finite arguments. Euclid had employed the same method for proving his theorems on geometry two millennia before, but then Euclid's axioms, or 'postulates', were regarded as self-evident truths. In Hilbert's formalism, symbols are shorn of meaning and only behave according to the list of axioms. The symbols do not have to represent anything at all, and to make the point Hilbert suggested that the results of geometry could still be deduced by replacing the words 'points and lines' by 'tables and chairs'.

If we adopt this 'formalist' approach, we are called upon to imagine a computer programmed with the axioms and rules of logic deducing strings of theorems in a finite number of steps. All that matters is that the symbols work within the axioms and rules. In Euclid's geometry care was taken to define a point and a line – a point being that which has no part, a line being a breadth without length – but in the formalist mode, which has done much to shape modern mathematics, we are not required to define mathematical objects too closely. Hilbert had high hopes that formalism would provided a satisfactory model for mathematics.

Intuitionists and the incompleteness theorems

Opposition to the ideas of both Hilbert and Russell came from the Dutchman Luitzen E.J. Brouwer. He rejected the identification of mathematics with logic along the lines stated by Russell, and neither did he believe that we can know the objects of mathematics by setting down a list of axioms to which they should conform. Brouwer thought that mathematics had priority over logic and that mathematical objects were known through 'intuition'. In saying that mathematical objects are intuited, Brouwer and his followers showed they were sceptical about the handling of infinite sets. In particular, they rejected the 'law of the excluded middle', a principle of logic handed down from Ancient Greek times. This states that any proposition is either true or its negation is true. Formalists like Hilbert were horrified at the intuitionists' stance, realizing that many of the results of classical mathematics, safely deposited in the mathematical storehouse of knowledge, would have to be abandoned if the excluded middle principle were jettisoned. Gone would be wonderful theorems such as Euclid's for proving that the number of primes was infinite.

Hilbert's call for mathematics to be based on formal axiomatic systems received a mortal blow in 1931, when the Austrian mathematician and philosopher Kurt Gödel proved the 'incompleteness theorems'. Gödel proved that within a formal system that described the arithmetic of the natural numbers 1, 2, 3, 4, ... there were statements that could not be proved nor disproved. These statements thus remain undecidable, and the formal system incomplete. In a follow-up theorem, Gödel proved that if an axiomatic system is consistent, then its consistency cannot be proved from within the system.

> *'Gödel's theorem implies that pure mathematics is inexhaustible. No matter how many problems we solve, there will always be other problems that cannot be solved within the existing rules.'*
>
> FREEMAN DYSON

Gödel showed us that the idea of mathematical truth is fraught with difficulty, for there are results in formal arithmetic that cannot be proved in a finite number of computer-like steps. This is contrary to Hilbert's belief that true statements must be provable. One effect of Gödel's work is that we might start believing that some of the difficult problems in mathematics are simply irresolvable. But we also know that, though it might take hundreds of years, some thorny problems *do* get solved in the end: witness the proof of Fermat's last theorem (see *Is Mathematics Beautiful?*) and the Poincaré conjecture (see *What Shape is the Universe?*). Perhaps the true lesson here is that, in the end, the ingenious powers of the human mind can always surpass the computational capacity of a machine.

Mathematical truth — a contradiction?

So what of our book of mathematical truth? Gödel showed us it was not as secure as was once thought. In different language, the Hungarian-born philosopher Imre Lakatos argued that there is no such thing as certain mathematical knowledge; rather, the theorems continually change as the concepts to which they apply change their meaning.

There has been, in the 20th century, a revision of what we mean when we ask 'Is mathematics true?' In the words of Richard Courant and Herbert Robbins in their landmark book *What is Mathematics?*, 'Renouncing the goal of comprehending the "thing in itself", of knowing the "ultimate truth", was a great step forward.' Gödel showed that logical perfection is illusory, but in some ways that freed mathematicians to explore. The sense that truth, like beauty, may be in the eye of the beholder has its liberating aspect, and it could be said that mathematicians today pursue truth as they see it.

IS THERE ANYTHING LEFT TO SOLVE?

The great unsolved problems and the future of mathematics

*T*he cut and dried answers that receive ticks or crosses in a schoolchild's exercise book might suggest that mathematics as a whole is bounded and fixed. Of course, this is very far from the truth. There remain some famous and still-baffling conundrums, which have obsessed and frustrated generations of would-be solvers. And each problem that is solved potentially raises hundreds more, keeping mathematics a living and ever changing organism.

Mathematics thrives on unsolved problems, and there are many. Some have been around for centuries, but newly emerging theories suggest new challenges. Mathematicians periodically produce 'shopping lists' of key unsolved problems, the best-known example being David Hilbert's '23 Problems', issued in 1900 as a challenge for the new century's mathematicians. Some of these challenges were specific problems, while others were open-ended programmes. But, given Hilbert's high status in the mathematical world, they attracted a good deal of attention, and all 23 tasks have illuminated the mathematical landscape over the course of a century and more. They became a focus for the collective action of the world's best mathematicians.

By way of incentive, some lucrative rewards emerged for those mathematicians able to demonstrate proofs or solutions of selected tough problems. One of the first was the Wolfskehl

Prize of 100,000 marks, left by German tycoon Paul Wolfskehl for the first valid proof of Fermat's last theorem (see *Is Mathematics Beautiful?*). That particular theorem was finally proved true in 1994, by Sir Andrew Wiles. Perhaps the most well-known prizes in the English-speaking world are those offered by the Clay Institute, a private foundation dedicated to the spread of mathematical knowledge. On offer is $1 million for breakthroughs in each of seven areas adjudged significant for the advance of mathematics in the 21st century. Of these challenges, the Poincaré conjecture, concerning the topology of geometric objects, has now been proved (see *What Shape Is the Universe?*).

Among today's surviving conundrums, each mathematician has his or her favourite puzzle. But there is a general consensus about the cream of them, and that group includes the Goldbach conjecture, the Riemann hypothesis, the Navier–Stokes equations and the succinctly phrased $P = NP$. They are worth looking at in turn.

The Goldbach conjecture

The Goldbach conjecture was formulated in the mid 18th century and named after Christian Goldbach. It gets a mention in Hilbert's list and has resisted all attempts at proof. Number theory is at the core of all mathematics, and it generates some of the most difficult problems, of which the Goldbach conjecture is emblematic. The problems of number theory have the virtue of being easy to state but difficult to prove.

As with many problems of this type, the Goldbach conjecture involves expressing numbers as the sum of prime numbers:

Every *even* whole number greater than or equal to 4 can be written as the addition of two prime numbers.

To test it, we can choose an even number at random, say 407,308, and verify it. Indeed we find that 407,308 is the sum of two prime numbers: 17 + 407,291. No one has yet managed to find an exception to Goldbach's conjecture, but no one has managed to prove it either, so its conjectural quality remains.

Numerical results have shown it to be true for all numbers up to 10^{18}, so there is no point in trying to find an exception among anything smaller than that.

The Riemann hypothesis

The Goldbach conjecture is a serious challenge. But the celebrated Riemann hypothesis has real mathematical weight. This, the eighth problem on Hilbert's list, is the most eagerly sought proof in all of mathematics, and fame awaits anyone who succeeds – as well as a useful $1 million. So many results in number theory hinge on its truth, and it is not short of attention from mathematical detectives.

The Riemann hypothesis can be framed in terms of what is known as the 'Riemann zeta function'. The hypothesis originated with Bernhard Riemann in the mid 19th century and is named after the sixth letter ζ (zeta) of the Greek alphabet. The function is expressed as:

$$\zeta(s) = 1 + \frac{1}{2^s} + \frac{1}{3^s} + \frac{1}{4^s} + \cdots$$

It can be calculated by choosing a value of s, so for example if $s = 2$, the function gives the value:

$$\zeta(2) = 1 + \frac{1}{2^2} + \frac{1}{3^2} + \frac{1}{4^2} + \cdots$$

There seems no reason why this series should be linked with the constant π, but a famous result of Leonhard Euler's is that the value of $\zeta(2)$ is $\frac{\pi^2}{6}$.

One of the most remarkable things about the Riemann zeta function is its connection with prime numbers. Euler discovered that the function was equal to the multiplication of all numbers of the form

$$\frac{p^s}{p^s - 1}$$

where p is a prime number. So, for example:

$$\zeta(2) = \frac{2^2}{2^2 - 1} \times \frac{3^2}{3^2 - 1} \times \frac{5^2}{5^2 - 1} \times \dots = \frac{4}{3} \times \frac{9}{8} \times \frac{25}{24} \times \dots$$

In his famous 1859 paper 'On the Number of Primes Less than a Given Magnitude', Riemann introduced the possibility of s being a 'complex number', that is, a two-dimensional number (see *Which Are the Strangest Numbers?*).

A complex number has the form $a + bi$ for numbers a, b (the i is the 'imaginary'; see *Are Imaginary Numbers Truly Imaginary?*), and Riemann set out to find those that made $\zeta(s) = 0$. The first three values are approximately:

$$\frac{1}{2} + 14i, \qquad \frac{1}{2} + 21i, \qquad \frac{1}{2} + 25i$$

There are an infinite number of these values, but the ones that have been calculated – and billions have – all possess $\frac{1}{2}$ as the first component. The Riemann hypothesis is that this propery is the case for *all* values. So far, it has been proved that an infinite number of values have this property. The problem is that, logically, that leaves the possibility that an infinite number of solutions *do not* have this property.

Riemann showed that the conjecture is connected to the way the prime numbers are distributed along the number line. The prime numbers are dense within the first hundred numbers (say), but in the long expanse of the number line there are huge gaps between some prime numbers. The Riemann conjecture holds the key to knowing the exact details of this distribution.

The Navier–Stokes equations

When we fly in aircraft or sail on ships, we may be unlucky enough to experience air or water turbulence. Water and air are both classified as fluids, the properties of which mathematicians and scientists have studied for centuries. In the first half of the 19th century, the French physicist Claude-Louis Navier and the Irish-born mathematician George Gabriel Stokes were working independently on fluids. Today, turbulence is mathematically modelled by the Navier–Stokes equations, named after the two men.

These equations are derived from the physical principles of the conservation of mass and of momentum. The trouble is, we cannot as yet solve them explicitly, and there is little in the way of mathematical theory that would serve to explain the nature of any solutions. If progress could be made, weather forecasting would become less of an art than a science, and other applications would benefit too. The curling motion of smoke and the pattern of flames from fire show similar physical features to the motion of air turbulence, and these might be modelled using the equations.

The solution to the Navier–Stokes equations is one of the holy grails of mathematics, and it is one of the six remaining Clay Institute challenges. Substantial insight into the nature of the equations, yielding a new deep mathematical theory of fluid dynamics is required for the award of the prize.

A good deal has been achieved with approximate numerical methods, and a whole new field known as computational fluid dynamics (CFD) is a highly active area of research. Coupled with CFD, computer graphics has enabled the visualization of turbulent flows to be modelled, especially useful in aircraft and ship design. Computer graphics can thus replace physical wind-tunnels with virtual ones, and these offer significant practical advantages, enabling, for example, airflow over a wing and water flow past a ship's hull to be simulated in real time. This means we can alter the speed of the virtual aircraft or ship and see the immediate effect on the design. It is not simple animation, because the flow pattern comes directly from the equations. As the speed of computers advances, these simulations will become even more realistic.

Does $P = NP$?

The mysterious-looking question of whether P is equal to NP is a problem that is situated where computer science meets mathematics. As part of 'computational complexity theory', it deals with limitations as to what computers are capable of doing.

It, too, is on the Clay Institute's 'most wanted' list of solutions and proofs. The question is a vital one in the 21st century, for it carries implications for computer security and the *algorithms* used in number theory. Computers can only work with algorithms, sequences of rules to be carried out, and while some algorithms take only micro-seconds to execute, others would, at current speeds, take billions of centuries.

The *efficiency* of an algorithm is the key idea here. A familiar task for a computer is sorting, placing names in alphabetical order or sorting numbers in ascending order. If human beings, for example, have to sort the numbers 5, 3, 4, 2, 1 in the correct order it is a fairly simple task: we *know* what order they should be in. But how would a computer tackle the task? One method used is the 'bubble' sort, an algorithm that looks at the adjacent *pairs* of numbers in a systematic way and either swaps them over if one is greater than the other, or leaves them alone if they are in the right order.

In the initial pass through the sequence the computer first swaps 5, 3 to produce 3, 5, 4, 2, 1. Looking at the second pair it swaps 5, 4 to give 3, 4, 5, 2, 1, and so on. After a first pass consisting of four such comparisons, the number 5 has 'bubbled' up to its correct position at the end of the sequence. In the subsequent pass the computer only has to deal with 3, 4, 2, 1. In total, the computer needs to make ten comparisons to get the right sequence.

We could broaden this out to say that, if we had n numbers to sort, we could count the number of comparisons in the same way. There would be a limit on the number of steps needed, for they would certainly amount to fewer than n^2 (just as 10 was less than $5^2 = 25$ in the example). Any algorithm where the number of steps is like a power of n is said to be solved in 'polynomial' time. Computers can easily handle problems of this sort, and these algorithms are the efficient ones.

Now let's consider how well a computer might tackle the famous 'travelling salesperson problem'. In this, there is a given number of cities and a variety of costs for travelling between cities. If a salesperson is given a route around all the cities, the question is whether there is a cheaper roundtrip. In this form of the problem, the input is n cities and the output will be a decision 'yes' or 'no'. How many computer steps are required to find a cheaper route? A brute-force approach might consider all possible routes.

If we have about 100 cities and a futuristic supercomputer with the capability of 10^{18} operations per second, this approach still requires something like 4,000 centuries to solve the problem. It would be wonderful, therefore, if there was an efficient algorithm for this problem, one that involves only the polynomial time that computers are comfortable with handling. If there was one with the efficiency of the bubble type, such supercomputers would dispose of the travelling salesperson problem in less than a micro-second.

With efficiency in mind, there are two important classes of problems to consider, P and NP problems:

- P is the class of problems which can be *solved* in polynomial time, such as the sorting problem.
- NP is the class of problems that can be *verified* in polynomial time. The travelling salesperson problem is one of these. Although it is extremely challenging to *find* a solution of the travelling salesperson problem, the calculation to *verify* whether the route was cheaper than the given route would be quick. It is the difference between the difficulty of finding the needle in the haystack and verifying that you have indeed found it.

Every problem in the class P is in the class NP, since finding the solution in polynomial time is its own verification. The burning question is: is every NP problem a class P problem? In other words, is it possible to find a polynomial-time algorithm for each NP problem? If it is, then $P = NP$. On the other hand, if we could find an NP problem and a proof that there could be no efficient algorithm for it, then P could not equal NP.

Here is where the travelling salesperson problem achieves importance. It is a hard problem, and all other problems in the *NP* class can be transformed into it. If we could find an efficient algorithm for just this *one* problem, it would follow that $P = NP$. The implication of that result would be that all *NP* problems could be solved efficiently.

One implication of a successful demonstration of $P = NP$ would be that we could find the prime factors of a large number by an efficient computer algorithm. Modern cryptology (see *Can We Create an Unbreakable Code?*) is based on the difficulty of this problem, and it provides the security on which electronic commerce and the integrity of computer networks is based. In other words, a successful proof would give hackers and cyber-thieves a field day. Luckily it is thought unlikely that machines can solve all problems efficiently, and opinion favours $P \neq NP$.

The future of mathematics

Continuing efforts to tame these big problems will doubtless occupy mathematicians for years to come. It may be that solutions and proofs emerge in a gradual way rather than explode upon the stage – that is often the way mathematics progresses, by refinement. Concepts are honed, labyrinthine calculations of yesteryear are jettisoned, arcane arguments become redundant, and the pathways to many areas are smoothed. As with any field of human thought, fashion and trends play their part too. Not all areas of mathematics can survive, and large fields of endeavour are swept away as theories are recast. Some of them, at best, become historical footnotes.

Does the recent past give clues as to the future? In at least one way, the answer is surely 'yes'. The second half of the 20th century saw the dawn of the computer age, with its profound impact on mathematics – and everything else. In the 1970s, the computer-generated proof of the 'four colour' map conjecture (see *Is There a Formula for Everything?*) was a sign of the times. Today's mathematicians can take for granted a computational capability that would have been unimaginable a few decades earlier. The mathematician is now very often more of a scientist in the laboratory. Ideas can be tested and mathematical experiments carried out. Not only arithmetical calculations, but also algebra can be 'computerized' without much trouble, and powerful graphics allow a wide range of geometrical shapes and surfaces to be seen. Mathematicians have always made physical models, but computer graphics do it better. We must assume, nevertheless, that the mathematicians of tomorrow, with their ever-more advanced computing resources, will regard our current capabilities as feeble.

There is no doubt, however, that the age of mathematics is upon us. Thirty years ago, mathematics was seen as applying to engineering and physics. Today it has spread its wings to new fields, and these in turn have demanded new mathematics. It is clear that computer science needs mathematics, but it was less predictable that chemistry, biology and psychology – to name but three examples – would also come to depend on mathematical input. Indeed, the entire mathematical field is truly vast, with scarcely a discipline untouched in some respect.

The professional mathematician is, consequently, now highly specialized, not merely in one branch of mathematics, but in one small corner of one area of that branch. This atomization is an inevitable consequence of the proliferation of knowledge, of theory, of technology and of mathematical applications. Several hundred years ago, one could have imagined gathering the world's leading mathematicians together and they would have talked animatedly (let's assume a shared tongue) and with mutual understanding, sharing the same intellectual horizons. A similar exercise today might produce a grouping of individuals profoundly well versed in their specialism but largely at a loss as to what their colleagues are doing. It is, perhaps, the price of progress.

Where does the future lie? Essentially, where it has always lain: with a curious mind, a pencil and a notepad (and maybe now a computer); and with an appetite to solve the as-yet unsolved. There is much still to do.

GLOSSARY

Algebra
Dealing with letters instead of numbers so as to extend arithmetic, algebra is now a general method applicable to all mathematics and its applications. The word 'algebra' derives from '*al-jabr*' used in an Arabic text of the ninth century AD.

Algorithm
A mathematical recipe; a set routine for solving a problem.

Argand Diagram
A visual method for displaying the two-dimensional plane of complex numbers.

Attractor
A point which attracts curves of motion in a dynamical system. In a pendulum which gradually comes to rest the curve of motion or trajectory is attracted to a single point where the displacement is 0 and the velocity is 0. For complicated dynamical systems the attractor may constitute a fractal – in which case it is called a 'strange attractor'.

Base
The basis of a number system. The Babylonians based their number system on 60, while the modern base is 10 (decimal).

Binary number
A number based on two symbols 0, 1. Binary numbers are fundamental for the operation of computers.

Butterfly effect
The effect of totally different trajectories generated by initial starting points very close together.

Cardinal number
The number of elements in a set. The cardinal number of $\{a, b, c, d, e\}$ is 5 but the notion of cardinal number can be extended to infinite sets.

Chaos theory
The theory of dynamical systems that appear random but have underlying regularity.

Clifford algebra
A type of n-dimensional algebra found to be useful in quantum mechanics. Due to mathematician W.K. Clifford.

Complex number
A number of the form $a + ib$ where a, b are ordinary real numbers and the imaginary i has the property $i^2 = -1$. The complex number $2 + 3i$ is an example.

Decimal system
Our commonly used number system based on the *ten* symbols, 0, 1, ..., 9.

Denominator
The bottom part of a fraction. In the fraction $\frac{3}{7}$, the number 7 is the denominator.

Differential equation
Equation involving concepts of Calculus, such as velocity, acceleration, and other rates of change. Prominent examples include Maxwell's equations and the Navier–Stokes equations.

Differentiation
A basic operation in Calculus which produces the derivative or rate of change. For an expression describing how distance depends on time, for example, the derivative represents the velocity. The derivative of the expression for velocity represents acceleration.

Duality
Interplay between two concepts, such as *point*, *line* in geometry; and between OR, AND in logic.

e
Euler's e is the second most important mathematical constant, after π. It occurs especially in problems of growth. The decimal value of e is approximately 2.7182818285...

Elliptical geometry
A non-Euclidean geometry in which the sum of the angles in a triangle is greater than 180°.

Empty set
The set with no objects in it. Traditionally denoted by ϕ, it is a useful concept in set theory.

Enigma
A sophisticated electrical-mechanical device used for encrypting messages during World War II.

Euclid's geometry
Traditional geometry as presented in Euclid's *Elements* of 300 BC. The sum of the angles of a triangle is equal to 180° in this geometry.

Exponent
A notation used in arithmetic. Multiplying a number by itself, 5×5 is written 5^2 with the exponent 2. The expression $5 \times 5 \times 5$ is written 5^3, and so on. The notation may be extended: for example, the number $5^{\frac{1}{2}}$ means the square root of 5. Equivalent terms are power and index.

Fano plane
A geometry consisting of seven points and seven lines.

Fermat's last theorem
This theorem states that there are no whole number values of x, y, z which make $x^n + y^n = z^n$ where n is a whole number greater than 2.

Fibonacci sequence
The sequence of whole numbers 1, 1, 2, 3, 5, 8, 13, 21, 34, 55, … where each term is the sum of the two preceding terms.

Fractal
A 'rough' shape which looks the same at any microscopic scale; a fractal has the property of self-similarity. The iconic example is the Mandelbrot Set.

Fraction
A whole number divided by another, for example $\frac{3}{7}$.

Geometry
Dealing with the properties of lines, shapes, and spaces, the subject was formalized in

Euclid's *Elements* in the third century BC. Geometry pervades all of mathematics and has now lost its restricted historical meaning.

Golden ratio
The ubiquitous 'golden number' φ (phi), whose value is:
$$\varphi = \frac{1 + \sqrt{5}}{2} = 1.6180339887\ldots$$

Group
A mathematical structure comprising elements which can be combined to give an element of the same kind, and which have certain properties: all elements have an inverse element, and the associative law $a \times (b \times c) = (a \times b) \times c$ is obeyed.

Hausdorff dimension
A way to measure dimension, which may give a fractional value. It has special applicability to the study of fractals.

Hexadecimal number
A number system of base 16 based on 16 symbols, 0, 1, 2, 3, 4, 5, 6, 7, 8, 9, A, B, C, D, E, and F. It is widely used in computing.

Hyperbolic geometry
A non-Euclidean geometry in which the sum of the angles in a triangle is less than 180°.

Imaginary numbers
Numbers involving the 'imaginary' $i = \sqrt{-1}$. They help form the complex numbers when combined with ordinary (or 'real') numbers.

Integration
A basic operation in Calculus that measures area. It can be shown to be the inverse operation of differentiation.

Inverse square law
Newton's law states that *any* two objects of masses m_1 and m_2 are attracted to each other with a force proportional to

$$\frac{m_1 \times m_2}{d^2}$$

Irrational numbers
Numbers which cannot be expressed as a
fraction (e.g. the square root of 2).

Limit
The limiting value of an expression in x, as
x gets closer and closer to some specified
number.

Logistic equation
An equation used to model population growth
from one year to the next: $x_{n+1} = r(1 - x_n)x_n$.

Manifold
A geometrical object which may be
complicated but, if a small part is viewed,
ordinary Euclidean geometry applies. A
doughnut is a surface with a hole but a small
part of it looks like an ordinary disc without
a hole.

Null hypothesis, H_0
The name for the hypothesis which
statisticians put to the test.

Numerator
The top part of a fraction. In the fraction $\frac{3}{7}$,
the number 3 is the numerator.

Octonions, Cayley numbers
Eight-dimensional 'imaginary' numbers
discovered in the 1840s.

One-to-one correspondence
The nature of the relationship when each
object in one set corresponds to exactly one
object in another set, and vice versa.

Pi, π
First appearing in connection with the
circle, this most famous constant occurs
in all areas of mathematics. The value of
π is approximately $\frac{22}{7}$ or, as a decimal,
3.1415926535...

Place-value system
The magnitude of a number depends on the
position of its digits. In 73, the place value of
7 means '7 tens' and of 3 means '3 units'.

Polyhedron
A solid shape with many faces. For example,
a tetrahedron has four triangular faces and a
cube has six square faces.

Prime number
A whole number which cannot be divided by
any number (apart from itself and 1) without
leaving a remainder. For example 7 is a prime
number, but 6 is not (because 6 can be divided
by 2 leaving no remainder); 2 is the first prime
number.

Probability
The measurement of chance on a scale from
0 (impossible) to 1 (certain).

Public key encryption
The system by which the keys used to
encrypt a message are made public. The
receiver who decrypts a message has other
information about the keys that is unattainable
by a potential hacker.

Pythagoras's theorem
If the sides of a right-angled triangle have
lengths x, y and z then $x^2 + y^2 = z^2$ where z is
the length of the longest side (the hypotenuse)
opposite the right angle.

Quaternions
Four-dimensional 'imaginary' numbers
discovered in the 1840s by W.R. Hamilton.

Rational numbers
Numbers that are either whole numbers or
fractions.

Remainder
If one whole number is divided by another
whole number, the number left over is the
remainder. The number 17 divided by 3 gives
5 with remainder 2.

Sequence
A row (possibly infinite) of numbers or
symbols.

Series
A row (possibly infinite) of numbers or symbols added together.

Set
A collection of objects. The set consisting of the names of some countries $C = \{$England, Russia, United States of America, India, Australia$\}$. The set of positive even numbers $E = \{2, 4, 6, 8, \ldots\}$

Sphere
An ordinary sphere is the 2-dimensional surface of a ball. It has the equation $x^2 + y^2 + z^2 = 1$ in three variables x, y, z. An n-dimensional sphere has a similar equation in $(n + 1)$ variables.

Square number
The result of multiplying a whole number by itself. The number 9 is a square number because $9 = 3 \times 3$. The square numbers are 1, 4, 9, 16, 25, 36, 49, 64, …

Square root
The number which, when multiplied by itself, equals a given number. For example, 3 is the square root of 9 because $3 \times 3 = 9$.

Squaring the circle
The problem of constructing a square with the same area as that of a given circle – using only a ruler for drawing straight lines and a pair of compasses for drawing circles. It cannot be done.

Symmetry
The regularity of a shape. If a shape can be rotated so that it fills its original imprint it is said to have rotational symmetry. A figure has mirror symmetry if its reflection fits its original imprint.

Tessellation
A way of filling a flat space with one or more given shapes so that all shapes fit exactly together. For example, it is possible to form a tessellation using only hexagons but not one using only pentagons.

Theorem
A term reserved for an established fact of some consequence.

Three-body problem
The attempt to analyse the motion of three bodies (such as the Moon, Earth, Sun) which are mutually attracted to each other.

Topology
Also known as 'rubber sheet' geometry, which concentrates on such properties of surfaces as those which are preserved when they are transformed by stretching and contracting, but not cutting. The measurement of lengths and angles has no part to play in topology as it does in ordinary geometry.

Transcendental number
A number that cannot be the solution of an algebraic equation such as $ax^2 + bx + c = 0$ where a, b, c are whole numbers, or ones involving higher powers of x.

Twin primes
Two prime numbers separated by at most one number: for example, the twins 11 and 13. It is not known whether there is an infinity of these twins.

Unit fractions
Fractions where the top (numerator) is equal to 1.

x–y axes
The idea due to René Descartes of plotting points having an x-coordinate (horizontal axis) and y-coordinate (vertical axis).

Zero-sum
Applies to game theory. What one player wins (+w) the other loses (–w) so the net sum (+w) + (–w) is zero.

INDEX

Quercus Publishing Plc
21 Bloomsbury Square, London, WC1A 2NS

First published in 2011

A catalogue record of this book is available from the British Library

ISBN
UK and associated territories: 978 1 84916 240 1
Canada: 978 1 84866 091 5

Designed by Patrick Nugent
All illustrations and diagrams by Patrick Nugent
Typeset by Lapiz Digital, India

Printed and bound in China

10 9 8 7 6 5 4 3 2 1